Lecture Notes in Control and Information Sciences

Edited by A. V. Balakrishnan and M. Thoma

Vol. 1: Distributed Parameter Systems: Modelling and Identification
Proceedings of the IFIP Working Conference, Rome, Italy, June 21–26, 1976
Edited by A. Ruberti
V, 458 pages. 1978

Vol. 2: New Trends in Systems Analysis
International Symposium, Versailles, December 13–17, 1976
Edited by A. Bensoussan and J. L. Lions
VII, 759 pages. 1977

Vol. 3: Differential Games and Applications
Proceedings of a Workshop, Enschede, Netherlands, March 16–25, 1977
Edited by P. Hagedorn, H. W. Knobloch, and G. J. Olsder
XII, 236 pages. 1977

Vol. 4: M. A. Crane, A. J. Lemoine
An Introduction to the Regenerative Method for Simulation Analysis
VII, 111 pages. 1977

Vol. 5: David J. Clements, Brian D. O. Anderson
Singular Optimal Control: The Linear Quadratic Problem
V, 93 pages. 1978

Vol. 6: Optimization Techniques
Proceedings of the 8th IFIP Conference on Optimization Techniques, Würzburg, September 5–9, 1977
Part 1
Edited by J. Stoer
XIII, 528 pages. 1978

Vol. 7: Optimization Techniques
Proceedings of the 8th IFIP Conference on Optimization Techniques, Würzburg, September 5–9, 1977
Part 2
Edited by J. Stoer
XIII, 512 pages. 1978

Vol. 8: R. F. Curtain, A. J. Pritchard
Infinite Dimensional Linear Systems Theory
VII, 298 pages. 1978

Vol. 9: Y. M. El-Fattah, C. Foulard
Learning Systems:
Decision, Simulation, and Control
VII, 119 pages. 1978

Vol. 10: J. M. Maciejowski
The Modelling of Systems with Small Observation Sets
VII, 241 pages. 1978

Vol. 11: Y. Sawaragi, T. Soeda, S. Omatu
Modelling, Estimation, and Their Applications for Distributed Parameter Systems
VI, 269 pages. 1978

Vol. 12: I. Postlethwaite, A. G. J. McFarlane
A Complex Variable Approach to the Analysis of Linear Multivariable Feedback Systems
IV, 177 pages. 1979

Vol. 13: E. D. Sontag
Polynomial Response Maps
VIII, 168 pages. 1979

Vol. 14: International Symposium on Systems Optimization and Analysis
Rocquentcourt, December 11–13, 1978;
IRIA LABORIA
Edited by A. Bensoussan and J. Lions
VIII, 332 pages. 1979

Vol. 15: Semi-Infinite Programming
Proceedings of a Workshop, Bad Honnef, August 30 – September 1, 1978
V, 180 pages. 1979

Vol. 16: Stochastic Control Theory and Stochastic Differential Systems
Proceedings of a Workshop of the „Sonderforschungsbereich 72 der Deutschen Forschungsgemeinschaft an der Universität Bonn"
which took place in January 1979 at Bad Honnef
VIII, 615 pages. 1979

Vol. 17: O. I. Franksen, P. Falster, F. J. Evans
Qualitative Aspects of Large Scale Systems
Developing Design Rules Using APL
XII, 119 pages. 1979

Vol. 18: Modelling and Optimization of Complex Systems
Proceedings of the IFIP-TC 7 Working Conference
Novosibirsk, USSR, 3–9 July, 1978
Edited by G. I. Marchuk
VI, 293 pages. 1979

Vol. 19: Global and Large Scale System Models
Proceedings of the Center for Advanced Studies (CAS)
International Summer Seminar
Dubrovnik, Yugoslavia, August 21–26, 1978
Edited by B. Lazarević
VIII, 232 pages. 1979

Vol. 20: B. Egardt
Stability of Adaptive Controllers
V, 158 pages. 1979

Vol. 21: Martin B. Zarrop
Optimal Experiment Design for Dynamic System Identification
X, 197 pages. 1979

For further listing of published volumes please turn over to inside of back cover.

Lecture Notes in Control and Information Sciences

Edited by A.V. Balakrishnan and M. Thoma

55

Ganti Prasada Rao

Piecewise Constant Orthogonal Functions and Their Application to Systems and Control

Springer-Verlag
Berlin Heidelberg GmbH 1983

Series Editors
A. V. Balakrishnan · M. Thoma

Advisory Board
L. D. Davisson · A. G. J. MacFarlane · H. Kwakernaak
J. L. Massey · Ya. Z. Tsypkin · A. J. Viterbi

Author
Professor Ganti Prasada Rao
Lehrstuhl für Elektrische Steuerung und Regelung
Ruhr-Universität Bochum
4630 Bochum 1
Federal Republic of Germany
(on Sabbatical Leave from the
Department of Electrical Engineering
Indian Institute of Technology
KHARAGPUR (WB) 721302
India)

AMS Subject Classifications (1980): 33-xx, 34-xx, 35-xx, 42-xx, 93-xx

ISBN 978-3-540-12556-3 ISBN 978-3-540-38648-3 (eBook)
DOI 10.1007/978-3-540-38648-3

Library of Congress Cataloging in Publication Data
Rao, Ganti Prasada, 1942-
Piecewise constant orthogonal functions and their
application to systems and control.
(Lecture notes in control and information sciences ; 55)
Bibliography: p.
Includes index.
1. System analysis. 2. Control theory. 3. Functions, Orthogonal.
I. Title. II. Series.
QA402.R35 1983. 515'.83 83-10416

Originally published by Springer-Verlag Berlin Heidelberg New York Tokyo in 1983

2061/3020-543210

To my parents

GANTI VENKATAPPADU
and
GANTI RAJESWARAMMA

with love and regards

About the author

GANTI PRASADA RAO was born in Seethanagaram, Andhra Pradesh, India, on the 25th August, 1942. He studied at the College of Engineering, Kakinada, and received the B.E. degree in Electrical Engineering from Andhra University, Waltair (India), in 1963 with first class and high honours. He received the M. Tech. (Control Systems Engineering) and Ph. D. Degrees in Electrical Engineering in 1965 and 1969 respectively, both from the Indian Institute of Technology (I.I.T.) Kharagpur.

From July 1969 to October 1971 he was with the Department of Electrical Engineering, PSG College of Technology, Coimbatore, as an Assistant Professor. In October 1971 he joined the Department of Electrical Engineering, I.I.T. Kharagpur as an Assistant Professor and became a Professor in May 1978. From May 1978 to August 1980, he was the Chairman of the Electrical Engineering Curriculum Development Cell at I.I.T. Kharagpur. From October 1975 to July 1976, he was with the Control Systems Center, UMIST, Manchester, England, as a Commonwealth Post doctoral Research Fellow. Presently he is with the Lehrstuhl für Elektrische Steuerung und Regelung, Ruhr-Universität Bochum, West Germany, as a Research Fellow of the Alexander von Humboldt Foundation.

He has research publications in the areas of mathematical instruments, time-varying systems, parametric phenomena, system identification, applications of Walsh and related piecewise constant functions, and fuzzy logic control.

Professor GANTI PRASADA RAO is a Senior Member of the IEEE and a Fellow of the Institution of Engineers (India).

PREFACE

Walsh functions appeared in mathematical literature sixty years ago. About thirty years later, extensive applications of Walsh functions in communications and signal processing have been suggested. It was in the last decade that their potential as attractive basis functions for signal characterization in control problems became evident. There have been parallel developments with block-pulse functions showing definite advantages in computational aspects.

This book attempts to unify Walsh functions, block-pulse functions, Haar functions etc. into a general class of piecewise constant orthogonal basis functions and presents a comprehensive account of the various applications of these functions in problems such as analysis, optimization and identification of continuous time dynamical systems. The overall treatment is in terms of general piecewise constant orthogonal basis functions although in some particular situations and illustrations, Walsh functions or block-pulse functions are prefered in view of simplicity.

The work of the author with his past students Drs. L. Sivakumar, T. Srinivasan and K.R. Palanisamy in India and the several interesting contributions by Professors C.F. Chen, S.G. Tzafestas, L.S. Shieh, Y.P. Shih and R.R. Mohler and their colleagues abroad, form the core of this book. It is hoped that this book would be of interest to both applied mathematicians and control engineers.

The author is grateful to many of his colleagues and students for their helpful roles in shaping this book. In particular, at IIT Kharagpur, Professor C.N. Kaul (Department of Mathematics) and Professor N. Kesavamurthy gave valuable comments. Professor H. Unbehauen at the Ruhr-Universität Bochum took much interest and encouraged the author in the preparation of the manuscript giving constructive suggestions at several stages. The author thanks Frau H. Hupp, Frau P. Kiesel and Frau E. Schmitt for their skill in transforming an untidy manuscript into a neat typescript. He is grateful to Frau M.-L. Schmücker and Frl. H. Vollbrecht for their excellent draftsmanship.

The author owes a debt of gratitude to the Alexander von Humboldt Foundation for a Research Fellowship supporting his visit to Bochum. It is during this visit that this book became a reality.

The author is very much indebted to his wife Meenakshi, and his children Nagalakshmi, Rajeswari and Venkatalakshminarayana for their patience, love, understanding and encouragement not only during the period of preparation of this book but also throughout the several years of his research.

Bochum,
January, 1983.

GANTI PRASADA RAO

CONTENTS

SPECIAL ACKNOWLEDGMENT

The author thanks the publishers of the following journals, for permission to include in this book, considerable parts from references as detailed below:

1. IEEE Proceedings, IEEE Transactions on Automatic Control: References W43b, W45, W47, W48, W49, and W51.

2. Proceedings IEE: References W44, W46, and W53.

3. Optimal Control Applications and Methods (John Wiley & Sons): Reference W37a.

4. International Journal of Systems Science (Taylor & Francis): Reference W50.

GANTI PRASADA RAO

I

PIECEWISE CONSTANT ORTHOGONAL BASIS FUNCTIONS

1.1 Introduction

Let t_z be a time interval. A real valued function $f(t)$ is said to be square-integrable on t_z if the Lebesgue integral

$$\int_{t_z} f^2(\mu)d\mu < +\infty .$$

A collection of all measurable square-integrable functions on t_z is denoted as $L_2^{t_z}$. If the scalar product of $f(t)$, $g(t) \in L_2^{t_z}$ is defined by

$$(f(t),g(t)) = \int_{t_z} f(\mu)g(\mu)d\mu \tag{1.1}$$

and the norm of $f(t)$ by

$$||f(t)|| = \sqrt{(f(t),f(t))} = \sqrt{\int_{t_z} f^2(\mu)d\mu} \tag{1.2}$$

then, it is well known [G2,G3], that $L_2^{t_z}$ is a real Hilbert space.

A set $\{\theta_i(t)\} \in L_2^{t_z}$, finite or countable, is called orthonormal if

$$(\theta_i(t),\theta_j(t)) = \begin{cases} 0, & i \neq j \\ 1, & i = j \end{cases} \tag{1.3}$$

i.e., it is orthogonal ($\theta_i \perp \theta_j$, $i \neq j$) and normal ($||\theta_i|| = 1$ for all i).

If $\hat{\theta}_i(t)$ are pairwise orthogonal on t_z, then

$$\theta_i(t) = \frac{\hat{\theta}_i(t)}{||\hat{\theta}_i(t)||} , \tag{1.4}$$

will define an orthonormal system.

If $\{\theta_i(t)\}$ is an orthonormal system of basis functions in t_z, then every $f(t) \in L_2^{t_z}$ can be represented as

$$f(t) = \sum_{i=1}^{\infty} f_i \theta_i(t). \tag{1.5}$$

The sum of the series is understood in the sense of convergence of the norm, and the coefficients f_i are uniquely given by

$$f_i = (f(t),\theta_i(t)), \quad i=1,2,\ldots \quad . \qquad (1.6)$$

$\{f_i\}$ is the set of Fourier coefficients which will also be referred to as the "spectrum" of f(t) with respect to $\{\theta_i(t)\}$. For its Fourier sum to coincide in $L_2^{t_z}$ with f(t), a necessary and sufficient condition is that

$$||f||^2 = \sum_{i=1}^{\infty} f_i^2 \quad \text{(Parsevals' condition)} \; . \qquad (1.7)$$

An orthonormal system satisfying Parsevals' condition is said to be closed or complete. The Fourier series of any $f(t) \in L_2^{t_z}$ with respect to such a system is convergent in the mean of order two to that function.

1.2. Systems of piecewise constant basis functions (PCBF) on normal interval

- Definitions

Let t_z be the normal interval $[0,1)$ and choose $\{\theta_i(t)\}$ as piecewise constant on each subinterval $[\frac{i-1}{m}, \frac{i}{m})$, i=1,2,...,m, with some integral value for m. The following systems of functions are important particular cases of PCBF.

1.3. Block-pulse functions (BPF)

An m-set of BPF is defined as

$$\beta_i(t) = \begin{cases} 1, & \frac{i-1}{m} \leq t < \frac{i}{m} \;, \text{ for all } i=1,2,\ldots,m \\ 0, & \text{otherwise} \;. \end{cases} \qquad (1.8)$$

A set of four BPF over $[0,1)$ is shown in Fig. 1.1. The functions are disjoint and orthogonal. That is,

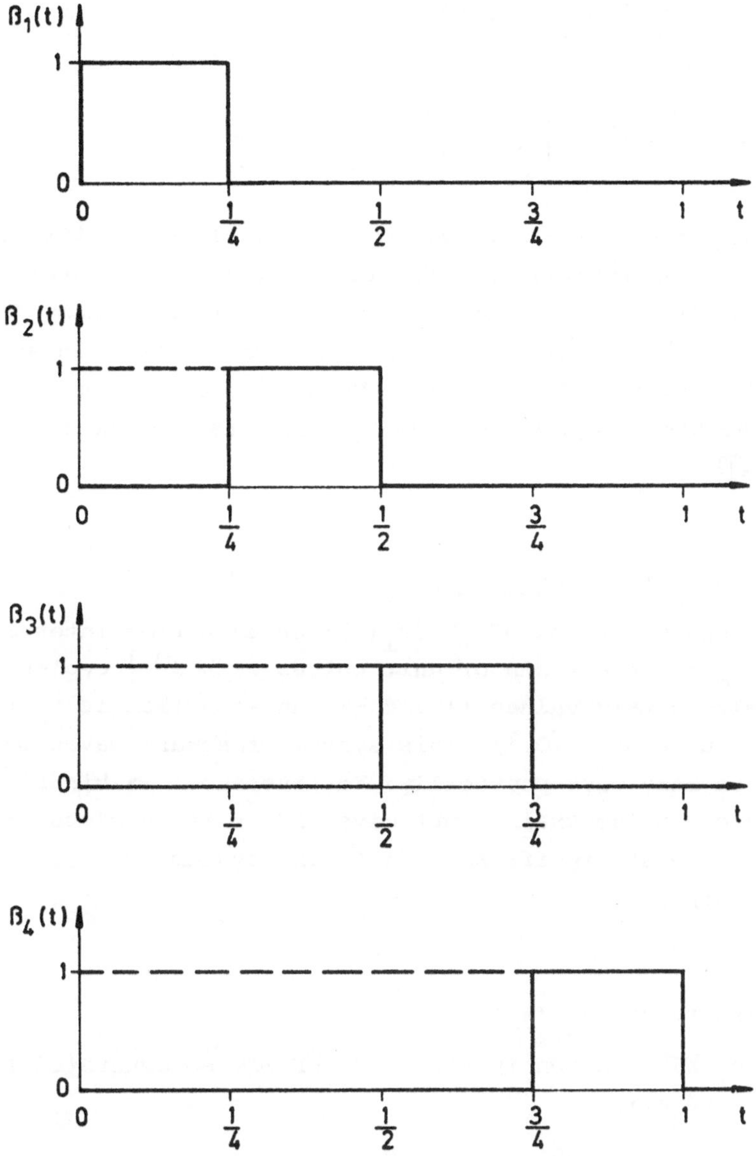

Fig. 1.1. A set of four BPF

$$\beta_i(t)\ \beta_j(t) = \begin{cases} 0\ , & i \neq j \\ \beta_i(t), & i=j \end{cases}$$

$$(\beta_i(t),\ \beta_j(t)) = \begin{cases} 0\ , & i \neq j \\ \frac{1}{m}\ , & i=j\ . \end{cases} \qquad (1.9)$$

The set $\{\beta_i(t)\}$ may be normalized to $\{\hat{\beta}_i(t)\}$ by letting $\hat{\beta}_i(t) = \sqrt{m}\ \beta_i(t)$ for all i. Thus $\{\hat{\beta}_i(t)\}$ is a disjoint orthonormal system.We can take BPF as the fundamental system to generate other systems of PCBF. The early forms of communication signals, such as those in Morse code, are block pulses. (cf. Harmuth [W26, W27] , Gopalsami and Deekshatulu [W25], Beauchamp [W4], Sannuti [W57], and Prasada Rao and Srinivasan [W53, W54]).

1.4. Rademacher functions (RF)

Fig. 1.2 shows a 4-set of RF $\{r_i(t)\}$ on unit time interval $[0,1)$. In general $r_m(t)$ is a train of unit pulses with 2^{m-1} cycles in $[0,1)$ taking alternately values +1 and -1. An exception is $r_o(t)$ which is the unit pulse over $[0,1)$. This system of square waves may be generated in many ways physically. For instance, a binary counter with a clock pulse train input gives RF at its various stages upto that of the most significant digit. The system $\{r_i(t)\}$ is orthonormal but not complete.

1.5. Walsh functions (WF)

A 4-set of WF is shown in Fig. 1.3. WF may be generated from RF using the relation

$$w_n(t) = [r_q(t)]^{d_q}[r_{q-1}(t)]^{d_{q-1}}[r_{q-2}(t)]^{d_{q-2}}\ \ldots\ldots, \qquad (1.10)$$

where $w_n(t)$ is the (n+1)-th member of $\{w_i(t)\}$ ordered in a particular way, and

$$q = \left|[\log_2 n]\right| + 1\ , \qquad (1.11)$$

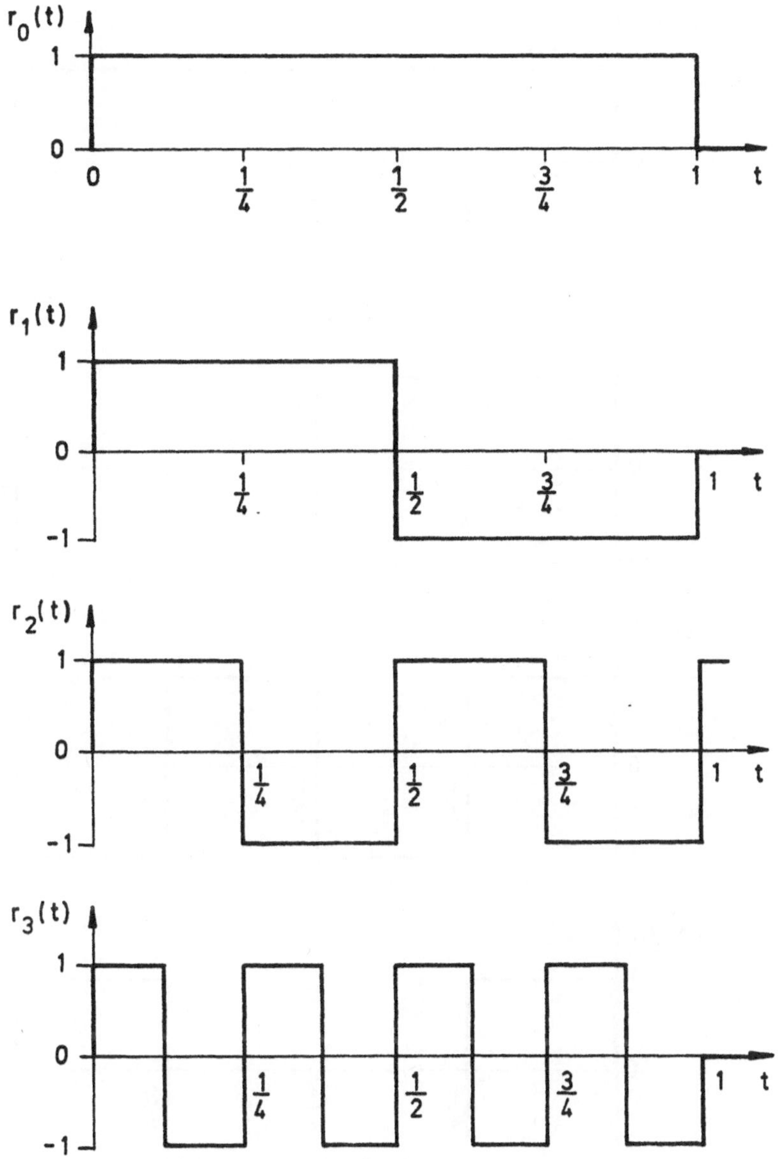

Fig. 1.2. A set of four Rademacher functions

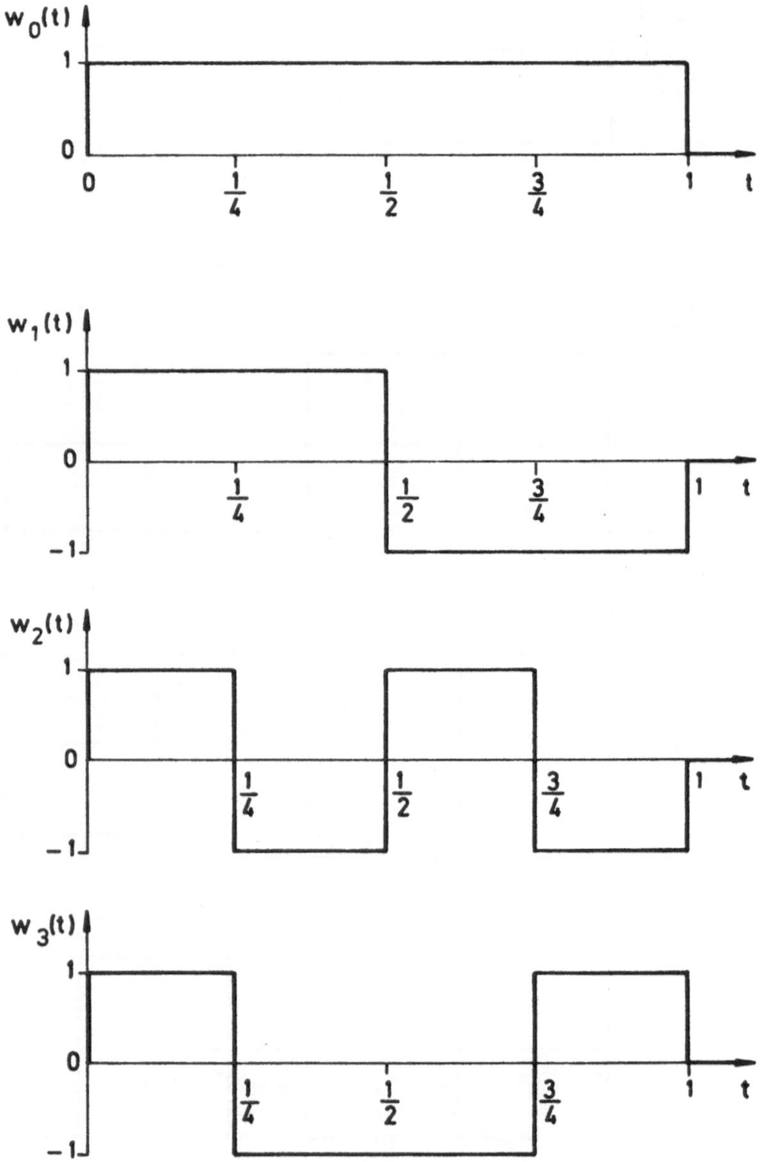

Fig. 1.3. A set of four Walsh functions

in which $\|[.]\|$ means taking the greatest integer of '.'. The digits 'd_k' (zeros or ones) belong to the binary form of 'n'. That is,

$$n = d_q 2^{q-1} + d_{q-1} 2^{q-2} + \ldots \quad . \tag{1.12}$$

There are many kinds of ordering of WF. We chose only one particular form called the Payley form here. In an m-set of WF, $m=2^k$, where k is a positive integer. Thus

$$\begin{aligned} w_o(t) &= r_o(t), \\ w_1(t) &= r_1(t), \\ w_2(t) &= r_2(t), \\ w_3(t) &= r_2(t)\, r_1(t), \\ \ldots & \quad \ldots \end{aligned} \tag{1.13}$$

and so on.

The system of WF is orthonormal and complete. For further information on WF such as generation, ordering, notations, and other aspects, the reader may see [W1-W4, W6, W9, W22, W24, W26, W27, W31, W58, W75].

1.6. Haar functions (HF)

Another complete system of orthonormal PCBF is that of Haar functions (HF) defined as an m-set as

$$\text{har}(0,0,t) = 1, \quad t \in [0,1),$$

$$\text{har}(j,n,t) = \begin{cases} 2^{j/2}, & \frac{n-1}{2^j} \le t < \frac{n-\frac{1}{2}}{2^j} \\ -2^{j/2}, & \frac{n-\frac{1}{2}}{2^j} \le t < \frac{n}{2^j} \\ 0, & \text{elsewhere}, \end{cases} \tag{1.14}$$

where $0 \le j \le \log_2 m$, $1 \le n \le 2^j$, $m=2^k$, and k is a positive integer. Fig. 1.4. shows a set of four HF on $[0,1)$, $1 \le n \le 2$, and m=4. The system is designated as $\{h_i(t)\}$ ordered as follows:

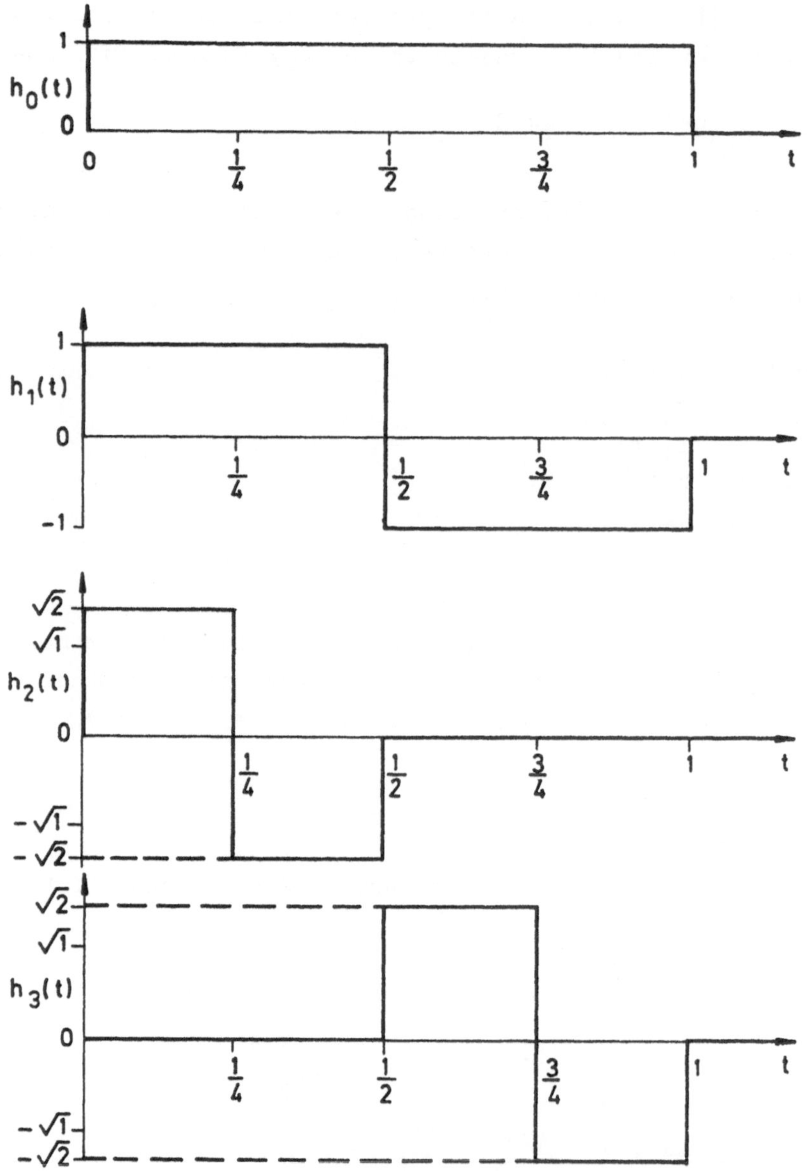

Fig. 1.4. A set of four Haar functions

$$\begin{aligned} h_o(t) &= har(0,0,t), \\ h_1(t) &= har(0,1,t), \\ h_2(t) &= har(1,1,t), \\ h_3(t) &= har(1,2,t). \end{aligned} \tag{1.15}$$

1.7. Relationships among the various systems of PCBF

Consider the first $m(m=2^k$, k an integer) terms in each of the series of PCBF defined above and write them concisely as m-vectors

$$\begin{aligned} \underline{\theta}(t) &= [\theta_1(t), \theta_2(t), \ldots, \theta_m(t)]^T, \\ \underline{\beta}(t) &= [\beta_1(t), \beta_2(t), \ldots, \beta_m(t)]^T, \\ \underline{r}(t) &= [r_o(t), r_1(t), \ldots, r_{m-1}(t)]^T, \\ \underline{w}(t) &= [w_o(t), w_1(t), \ldots, w_{m-1}(t)]^T, \\ \underline{h}(t) &= [h_o(t), h_1(t), \ldots, h_{m-1}(t)]^T. \end{aligned} \tag{1.16}$$

All the PCBF may be expressed as linear combinations of BPF. As a result of such a possibility it can be shown that

$$\begin{aligned} \underline{w}(t) &= \underline{T}_{BW}\,\underline{\beta}(t), \\ \underline{h}(t) &= \underline{T}_{BH}\,\underline{\beta}(t), \\ \underline{r}(t) &= \underline{T}_{BR}\,\underline{\beta}(t). \end{aligned} \tag{1.17}$$

The linear transformation matrices, in the case of m=4,are as follows:

$$\underline{T}_{BW} = \begin{bmatrix} 1 & 1 & 1 & 1 \\ 1 & 1 & -1 & -1 \\ 1 & -1 & 1 & -1 \\ 1 & -1 & -1 & 1 \end{bmatrix},$$

$$\underline{T}_{BH} = \begin{bmatrix} 1 & 1 & 1 & 1 \\ 1 & 1 & -1 & -1 \\ \sqrt{2} & -\sqrt{2} & 0 & 0 \\ 0 & 0 & \sqrt{2} & -\sqrt{2} \end{bmatrix},$$

$$\underline{T}_{BR} = \begin{bmatrix} 1 & 1 & 1 & 1 \\ 1 & 1 & -1 & -1 \\ 1 & -1 & 1 & -1 \\ \cdots & \cdots & \cdots & \cdots \end{bmatrix} .$$

$\underline{T}_{BW}$ and $\underline{T}_{BH}$ are constant invertible matrices, whereas $\underline{T}_{BR}$ is not even a square matrix. This suggests in simple terms that $\{r_i(t)\}$ is not complete. It is well known that the systems $\{w_i(t)\}$ and $\{h_i(t)\}$ are complete. The completeness of the system $\{\beta_i(t)\}$ has been first established by Prasada Rao and Srinivasan [W51].

Normalization may be done by letting $\hat{\beta}_i = \sqrt{m}\ \beta_i(t)$. With respect to the normalised system if we expand a function f(t), the i-th Fourier coefficient is given by

$$f_{\beta_i} = \frac{(f(t),\beta_i(t))}{\|\beta_i(t)\|} = \sqrt{m} \int_{\frac{i-1}{m}}^{\frac{i}{m}} f(t)dt . \tag{1.18}$$

By virtue of the mean value theorem, we have

$$f_{\beta_i}^2 = \frac{1}{m} f^2(t_i) \ , \ t_i \in \left[\frac{i-1}{m},\frac{i}{m}\right] . \tag{1.19}$$

As $m \to \infty$, $\frac{1}{m}$ may be replaced by an infinitesimally small interval dt. Consequently, integrating over $[0,1)$

$$\underset{m\to\infty}{\mathrm{Lt}} \sum_{i=1}^{m} f_{\beta_i}^2 = \|f(t)\|. \tag{1.20}$$

The BPF expansion of a function therefore satisfies (1.7) confirming the completeness of $\{\beta_i(t)\}$. This is further supported by the invertible transformations $\underline{T}_{BW}$ and $\underline{T}_{BH}$. These matrices could have been singular if the system $\{\beta_i(t)\}$ were not complete, since Walsh or Haar systems are.

1.8. Representation error in PCBF approximation

The representation error (or the residual error) when a differentiable function f(t) is represented in a series of PCBF over every subinterval $\left[\frac{i-1}{m},\frac{i}{m}\right)$ is (of Fig. 1.5)

$$e_i(t) = f_{\beta_i}\beta_i(t)-f(t),\ t\in\left[\frac{i-1}{m},\frac{i}{m}\right) \quad . \tag{1.21}$$

It can be shown that

$$\|e_i\|^2 = \frac{1}{12m^3}\left[\dot{f}\ (\xi_i)\right]^2 \qquad \xi_i\in\left[\frac{i-1}{m},\frac{i}{m}\right) \quad .$$

This leads to

$$\|e(t)\| \leq \frac{1}{2\sqrt{3}\ m}\ \sup_{t_z}(\dot{f}\ ) \ , \tag{1.22}$$

where

$$e(t) = \sum_{i=1}^{m} f_{\beta_i}\beta_i(t) - f(t).$$

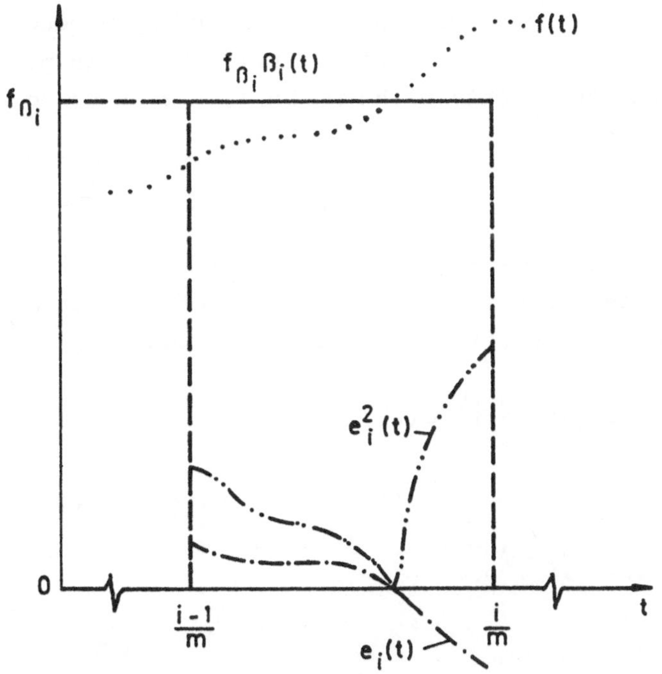

Fig. 1.5. Representation error over the i-th segment of a function represented in a series of PCBF

1.9. Choice of m in segmenting a signal for a specified error bound

If a signal $f(t) = A \sin \omega t$ has to be expanded in a series of PCBF with subinterval $1/m$, and if the allowable error $e(t)$ of representation is such that

$$\frac{\|e\|}{A} \leq \sigma_r ,$$

where σ_r is a constant specified error bound, then, from (1.22)

$$m \geq \frac{\omega}{2\sqrt{3}\sigma_r} .$$

In general, if a composite signal $f(t)$ has A_{max} and ω_{max} as the largest amplitude and frequency respectively in its sinusoidal spectrum, and then if

$$\frac{\|e\|}{A_{max}} \leq \sigma_r, \text{ then}$$

$$m \geq \frac{\omega_{max}}{2\sqrt{3}\sigma_r} .$$

In a PCBF expansion the signal is approximated by m constant valued segments by averaging over each of width $1/m$. In this process, all sinusoidal signal components with periods less than or equal to $1/m$ are totally blocked.

1.10. PCBF expansions of f(t)

The Fourier series with respect to $\{\theta_i(t)\}$ of a function $f(t)$ in t_z, truncated to retain the first m terms, may be compactly written as

$$f(t) \approx \sum_{i=1}^{m} f_i \theta_i(t) = \underline{f}^T \underline{\theta}(t) , \tag{1.23}$$

where

$$\underline{f} = [f_1, f_2, \ldots, f_m]^T .$$

For the particular cases of interest to us

$$f(t) \approx \underline{f}_\beta^T \underline{\beta}(t) \quad ,$$

$$\approx \underline{f}_w^T \underline{w}(t) \quad ,$$

$$\approx \underline{f}_h^T \underline{h}(t) \quad ,$$

where the vectors $\underline{f}_\beta$, $\underline{f}_w$ and $\underline{f}_h$ represent the related spectra. The spectra may be transformed from one basis to the other by using the linear transformations connecting the vectors $\underline{\beta}(t)$, $\underline{w}(t)$ and $\underline{h}(t)$. For instance,

$$\underline{f}_w^T \underline{w}(t) = \underline{f}_w^T \underline{T}_{BW} \underline{\beta}(t) = \underline{f}_\beta^T \underline{\beta}(t) \quad ,$$

implying that $\underline{f}_w^T = \underline{f}_\beta^T \underline{T}_{BW}^{-1}$.
Similarly,

$$\underline{f}_h^T = \underline{f}_\beta^T \underline{T}_{BH}^{-1} \quad , \text{ and so on.}$$

The system of RF being incomplete, we will not attempt any further discussions on function approximation with the Rademacher system.

1.11. Multidimensional PCBF and expansions [W55]

All the systems of functions discussed so far in single dimension of time may be straightaway generalised for multidimensional situations. To illustrate the procedure, let us consider the case of BPF in n dimensions.
A system of n-dimensional BPF is defined as

$$\{ \beta_{i_1, i_2, \ldots, i_n} (\kappa_1, \kappa_2, \ldots, \kappa_n) \}$$

over a region $\kappa_k \in [0,1)$, $k=1,2,\ldots,n$,
to form an orthogonal basis for approximating a multidimensional

square-integrable function $f(\kappa_1,\kappa_2,\ldots,\kappa_n)$. For each set of integers $\{i_1,i_2,\ldots,i_n\}$,

$$\beta_{i_1,i_2,\ldots,i_n}(\kappa_1,\kappa_2,\ldots,\kappa_n) = \begin{cases} 1 & \text{for } \frac{i_k-1}{m_k} \leq \kappa_k < \frac{i_k}{m_k} \\ 0 & \text{otherwise} \end{cases}$$

$$k=1,2,\ldots,n. \tag{1.25}$$

A multidimensional square-integrable function in the region $\kappa_k \in [0,1)$, $k=1,2,\ldots,n$, may be expanded as

$$f(\kappa_1,\kappa_2,\ldots,\kappa_n) \approx \sum_{i_1=1}^{m_1} \sum_{i_2=1}^{m_2} \cdots \sum_{i_n=1}^{m_n} f_{\beta_{i_1,i_2,\ldots,i_n}} \beta_{i_1,i_2,\ldots,i_n}(\kappa_1,\kappa_2,\ldots,\kappa_n) \;, \tag{1.26}$$

where m_k denotes the number of segments on the unit interval along the coordinate axis of the Cartesian direction κ_k. The set of BPF may be written as a vector $\underline{\beta}(\kappa_1,\kappa_2,\ldots,\kappa_n)$ of dimension $\prod_{k=1}^{n} m_k$. For instance when n=2 ,

$$\underline{\beta}(\kappa_1,\kappa_2) = [\beta_{1,1}\ldots\beta_{1,m_2},\beta_{2,1}\ldots\beta_{2,m_2},\ldots,\beta_{m_1,1}\ldots\beta_{m_1,m_2}]^T.$$

The multidimensional BPF may be separated, i.e.,

$$\beta_{i_1,i_2,\ldots,i_n}(\kappa_1,\kappa_2,\ldots\kappa_n) = \prod_{k=1}^{n} \beta_{i_k}(\kappa_k) \;. \tag{1.27}$$

The Fourier coefficients are obtained from the scalar products as

$$f_{\beta_{i_1,i_2,\ldots,i_n}} = (f,\beta_{i_1,i_2,\ldots,i_n}(\kappa_1,\kappa_2,\ldots,\kappa_n))$$

$$= \prod_{i=1}^{n} m_i \int_{\frac{i_1-1}{m_1}}^{\frac{i_1}{m_1}} \int_{\frac{i_2-1}{m_2}}^{\frac{i_2}{m_2}} \cdots \int_{\frac{i_n-1}{m_n}}^{\frac{i_n}{m_n}} f(\kappa_1,\kappa_2,\ldots,\kappa_n)d\kappa_1 d\kappa_2\ldots d$$

$$\kappa_k \in \left[\frac{i_k-1}{m_k}, \frac{i_k}{m_k}\right] \;. \tag{1.28}$$

The existing relationships among the PCBF in one dimension may be extended to the case of several variables. We list here these relations, for instance, when n=2. Each set of $(m_1 x m_2)$ basis functions may be written in the form of $m_1 m_2$-vectors

$$\underline{\theta}(\kappa_1,\kappa_2) = [\theta_{1,1} \ldots \theta_{1,m_2},\ldots,\theta_{m_1,1} \ldots \theta_{m_1,m_2}]^T,$$

$$\underline{\beta}(\kappa_1,\kappa_2) = [\beta_{1,1} \ldots \beta_{1,m_2},\ldots,\beta_{m_1,1} \ldots \beta_{m_1,m_2}]^T, \tag{1.29}$$

$$\underline{w}(\kappa_1,\kappa_2) = [w_{o,o} \ldots w_{o,m_2-1},\ldots,w_{m_1-1,1} \ldots w_{m_1-1,m_2-1}]^T,$$

$$\underline{h}(\kappa_1,\kappa_2) = [h_{o,o},\ldots,h_{o},m_{2}-1,\ldots,h_{m_1-1,1} \ldots h_{m_1-1,m_2-1}]^T.$$

Then,

$$\underline{w}(\kappa_1,\kappa_2) = \underline{I}_{BW}\underline{\beta}(\kappa_1,\kappa_2),$$

$$\underline{h}(\kappa_1,\kappa_2) = \underline{I}_{BH}\underline{\beta}(\kappa_1,\kappa_2), \tag{1.30}$$

where

$$\underline{I}_{BW} = \underset{(m_1 x m_1)}{\underline{T}_{BW}} \otimes \underline{I}_{(m_2 x m_2)},$$

and

$$\underline{I}_{BH} = \underset{(m_1 x m_1)}{\underline{T}_{BH}} \otimes \underline{I}_{(m_2 x m_2)},$$

with the Kronecker product defined as

$$\underline{A} \otimes \underline{B} = \{a_{ij}\underline{B}\}. \tag{1.31}$$

Each of the two dimensional PCBF is separable, i.e., $\theta_{i,j}(\kappa_1,\kappa_2) = \theta_i(\kappa_1)\theta_j(\kappa_2)$. Figures 1.6 - 1.8 show (4x4)-sets of the various two-dimensional PCBF.

Of the three complete systems of PCBF, namely, BPF, WF and HF, the BPF are simple and easy to visualise. BPF happen to be the fundamental

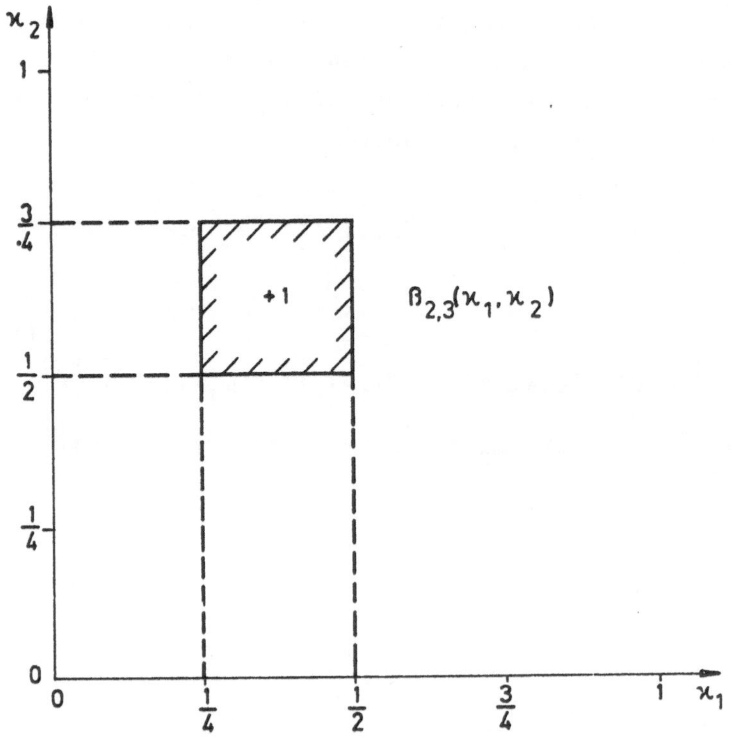

Fig. 1.6. The element $\beta_{2,3}$ of a 4x4 set of two-dimensional block pulse functions

system from which other systems can be generated through appropriate linear transformations. WF have been extensively used in signal processing for communication. Kitai [W31] gave a method of relating Walsh and sinusoidal spectra. With the relations among the various PCBF systems discussed here, it is possible to relate other PCBF spectra with the corresponding sinusoidal spectra. Two-dimensional WF have been used in picture processing, pattern analysis and other related fields. Parallel techniques with the other PCBF may be developed straightaway, bringing in the necessary modifications in the hardware. Walsh function generators and spectral analysers are described in [W69, W24].

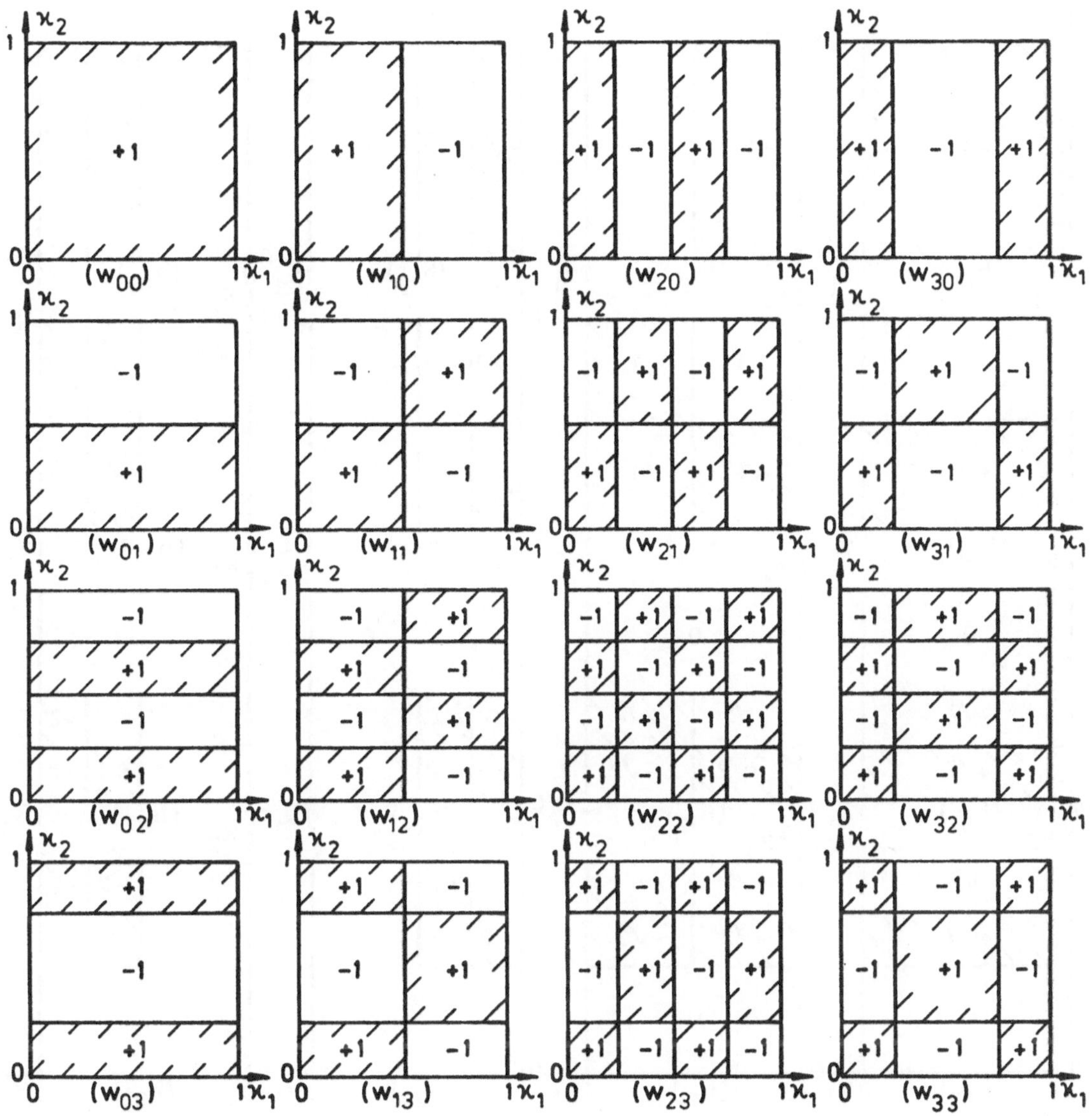

Fig. 1.7. A 4x4 set of two-dimensional Walsh functions

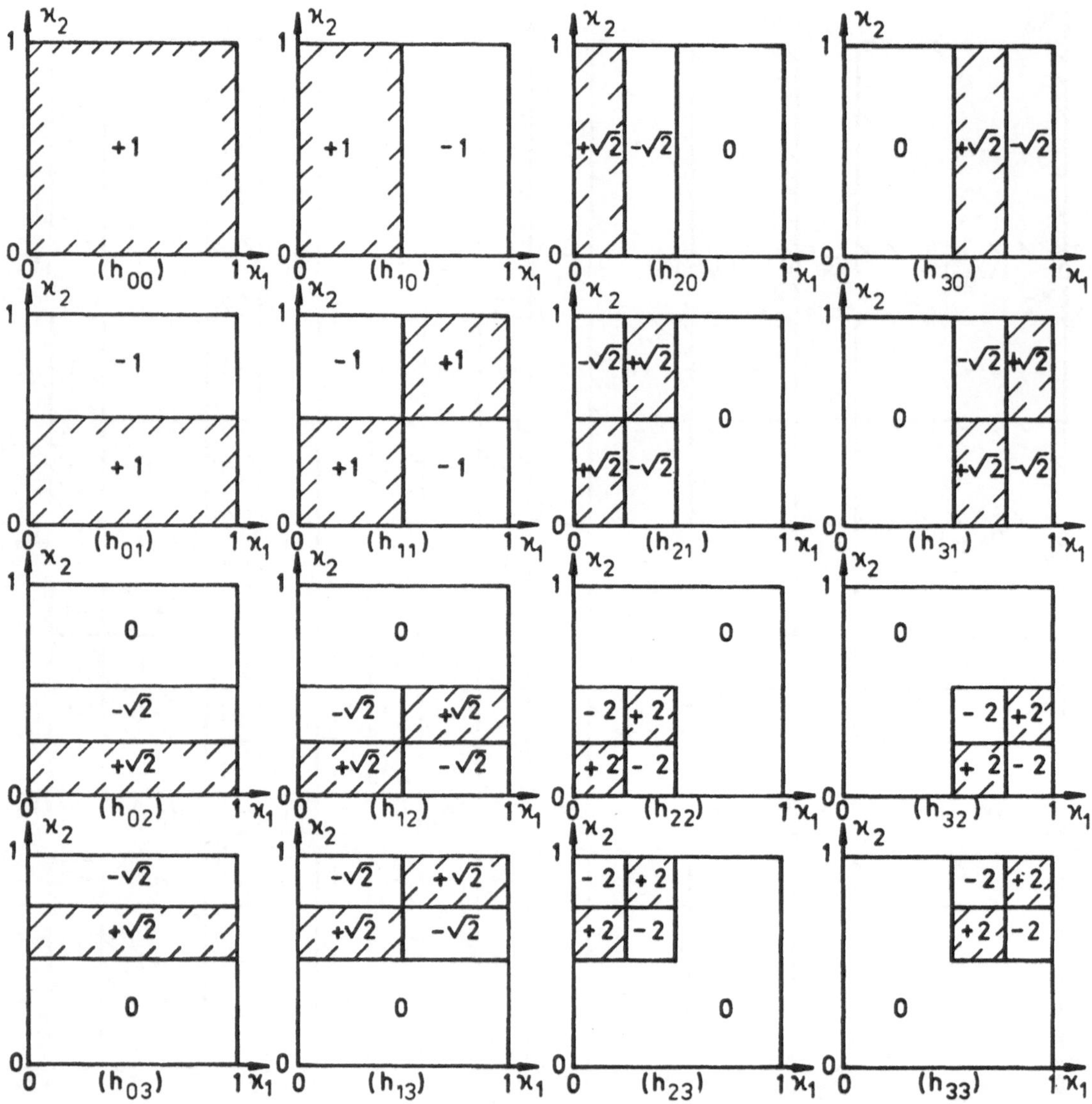

Fig. 1.8. A 4x4 set of two-dimensional Haar functions

1.12 A simple Walsh spectral analyser (WSA)

Fig. 1.9 shows a simple Walsh spectral analyser (WSA) as developed by Venkatesam and Prasada Rao [W74]. It consists of a Walsh function generator (WFG) designed to generate the first sixteen Walsh functions whose period may be chosen as desired. The Walsh functions obtained from the WFG are used to compute the spectrum in the next stage.

A standard positive-edge-triggered binary counter gives the first four RF. Multiplication of RF according to (1.11) is achieved by 'modulo 2 addition'. A set of 'exclusive OR' gates with inverters in cascade where necessary, is used to develop the set of WF. The outputs of the gates are Paley ordered. The outputs are at levels '0' and '+1'. The bias is removed by a level shifting network. Two transistors are used in tandem in this level shifting device. The circuit is designed first for any standard emitter current through

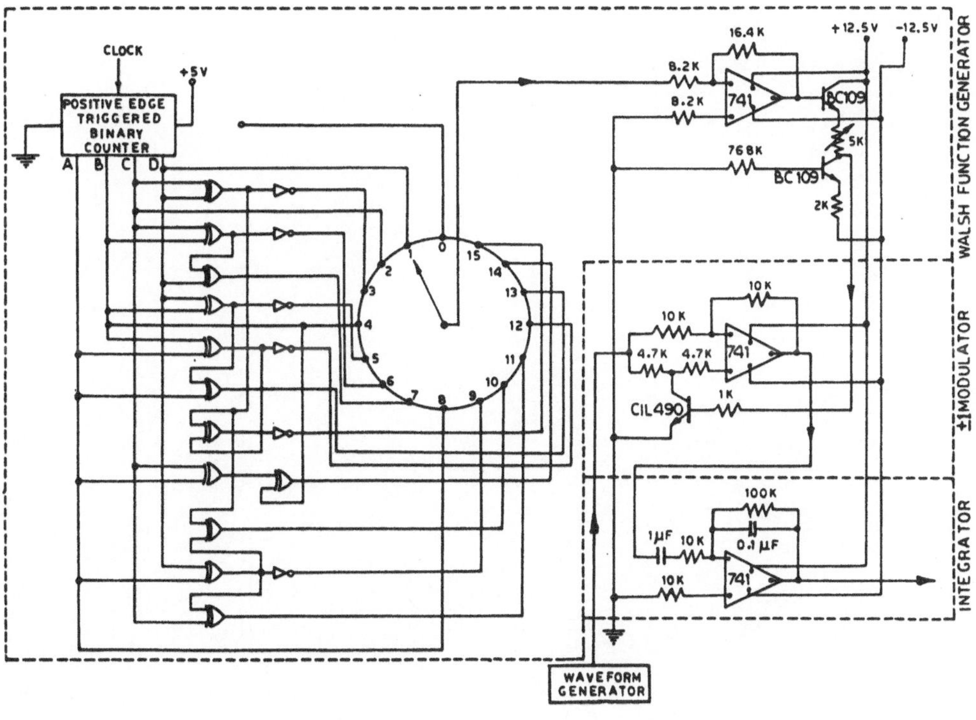

Fig. 1.9. A Walsh spectral analyser

the two transistors. The resistors are accordingly chosen to deliver symmetric biasfree output. A potentiometer facilitates adjustment.

<u>The ±1 modulator</u>

This is essentially a ±1 multiplier realised through a transistor switch. The circuit has an operational amplifier with unity gain. The inverting (or noninverting) terminal is grounded through a transistor. The base of this transistor receives a Walsh function signal and becomes forward or reverse biased depending on the instanteneous level of the Walsh function. Consequently, the inverting (or noninverting) terminal of the operational amplifier is grounded through the transistor or connected to the input. The result is a signal modulated by a Walsh function. The circuit elements of the integrator should be chosen according to the bandwidth requirements of signal analysis.

OPERATIONS ON SQUARE INTEGRABLE FUNCTIONS IN TERMS OF PCBF SPECTRA

2.1. Introduction

Consider two square-integrable functions f(t) and g(t) on the interval $[0,1)$. Let $\underline{f}$ and $\underline{g}$ be the m-vectors of spectral components of f(t) and g(t) respectively with respect to the general PCBF $\{\theta_i(t)\}$. Certain important time domain operations on the functions and some mutual operations between two given functions may be performed in terms of their PCBF spectra.

2.1.1. Addition (or subtraction)

If $c(t) = f(t) \pm g(t)$,
then,

$$\underline{c} = \underline{f} \pm \underline{g},$$

where

$$\underline{c} = [c_1\ c_2\ \dots\ c_m]^T,$$

$$c_i = (c(t),\ \theta_i(t)).$$

2.1.2. Multiplication (or division)

If $c(t) = f(t)\ g(t)$,
then,

$$\underline{c} = \underline{f} \boxed{\times} \underline{g},$$

and if

$$c(t) = f(t)/g(t), \quad g(t) \neq 0, \text{ for } t \in t_z,$$

$$\underline{c} = \underline{f} \boxed{\div} \underline{g}.$$

The symbols $\boxed{\times}$ and $\boxed{\div}$ represent appropriate operators which will be detailed with reference to each system of PCBF.

2.1.3. Integration with respect to 't'

Consider

$$\int_0^t f(\lambda)d\lambda \;, \; t\epsilon t_Z \,.$$

If $f(t) \approx \underline{f}^T\underline{\theta}(t)$, an m-term aaproximation, the integral may be written as

$$\int_0^t \underline{f}^T \; \underline{\theta}(\lambda)d\lambda \,.$$

We now integrate each term in the column vector $\underline{\theta}(t)$ and expand the result in a series of $\{\theta_i(t)\}$ again. We get

$$\int_0^t \underline{f}^T \; \underline{\theta}(\lambda)d\lambda \approx \underline{f}^T \; \underline{E\theta}(t), \tag{2.1}$$

where $\underline{E}$ is a constant invertible matrix. The integration is approximately achieved by premultiplying the spectral vector with the $\underline{E}$ matrix. The result is of considerable importance to us in reducing the calculus of continuous dynamical systems to an approximate (in the sense of least squares) matrix algebra.

2.1.4. Differentiation

In view of the fact that

$$\int_0^t \dot{f}(\lambda)d\lambda = f(t) - f(0),$$

by virtue of the approximation discussed just above, we get

$$\underline{f}_d = \underline{E}^{-1}(\underline{f} - f(0)\underline{u}_p) \,, \tag{2.2}$$

where $\underline{f}_d$ and $\underline{u}_p$ represent the PCBF spectral vectors of f(t) and unit step function respectively.

2.1.5. Time delay

Given a constant τ and a function whose PCBF expansion is $\underline{f}^T\underline{\theta}(t)$ over an interval $[0,1)$, we proceed as follows to obtain the spectrum of the delayed function $f(t-\tau)$.

1) Obtain the PCBF expansion of f(t) over the previous interval $[-1,0)$ denoted by $\overleftarrow{\underline{f}}^T \overleftarrow{\underline{\theta}}(t)$, where the '←' signifies association with the previous interval. $\overleftarrow{\underline{f}}^T$ corresponds to the carry over spectrum. $\overleftarrow{\underline{\theta}}(t)$ is a vector of PCBF defined on $[-1,0)$.

2) Consider $\underline{\theta}(t-\tau)$. Each element of this vector is a PCBF delayed by a known constant τ. Expand each $\theta_i(t-\tau)$ in an m-set of PCBF again.

3) Express the result in the form

$$\underline{\theta}(t-\tau) \approx \underline{D}\,\underline{\theta}(t) + \text{Spectrum carried over from } [-1,0). \quad (2.3)$$

$\underline{D}$ is the operational matrix for delay with respect to $\underline{\theta}(t)$.

2.1.6. Function with a stretched argument

If the PCBF expansion of a function on $[0,1)$ is $\underline{f}^T\,\underline{\theta}(t)$, then that of the stretched function $f(t/\lambda)$ for a known constant $\lambda>1$, may on the lines described above, be written as $\underline{f}^T\underline{S}\underline{\theta}(t)$. To obtain the operational matrix $\underline{S}$, for stretch, we stretch each element $\theta_i(t)$ of $\underline{\theta}(t)$ and rerepresent the result in terms of $\underline{\theta}(t)$.

2.1.7. Operational matrices for partial calculus for multidimensional functions

For a vector of 2-dimensional spectral components $\underline{\theta}(\kappa_1,\kappa_2)$, we get the following results:

$$\int_0^{\kappa_1} \underline{\theta}(\kappa_1,\kappa_2)d\kappa_1 = \underline{E}_{\kappa_1}\,\underline{\theta}(\kappa_1,\kappa_2)$$

$$= [\underline{E}_{(m_1 x m_1)} \otimes \underline{I}_{(m_2 x m_2)}]\,\underline{\theta}(\kappa_1,\kappa_2), \quad (2.4a)$$

$$\int_0^{\kappa_2} \underline{\theta}(\kappa_1,\kappa_2)d\kappa_2 = \underline{E}_{\kappa_2}\,\underline{\theta}(\kappa_1,\kappa_2)$$

$$= [\underline{I}_{(m_1 x m_1)} \otimes \underline{E}_{(m_2 x m_2)}]\,\underline{\theta}(\kappa_1,\kappa_2), \quad (2.4b)$$

and

$$\int_0^{\kappa_1}\int_0^{\kappa_2} \underline{\Theta}(\kappa_1,\kappa_2)d\kappa_1 d\kappa_2 = \underline{E}_{\kappa_1\kappa_2}\underline{\Theta}(\kappa_1,\kappa_2) = \underline{E}_{\kappa_2\kappa_1}\underline{\Theta}(\kappa_1,\kappa_2)$$

$$= [\underline{E}_{(m_1 x m_1)} \otimes \underline{E}_{(m_2 x m_2)}]\underline{\Theta}(\kappa_1,\kappa_2) \; . \qquad (2.4c)$$

In the general case of n dimensions,

$$\int_0^{\kappa_k} \underline{\Theta}(\kappa_1,\kappa_2,\ldots,\kappa_n)d\kappa_k = \underline{E}_{\kappa_k}\underline{\Theta}(\kappa_1,\kappa_2,\ldots\kappa_n),$$

$$\underline{E}_{\kappa_k} = [\underline{I}_{(m_1 x m_1)} \otimes \underline{I}_{(m_2 x m_2)} \otimes \ldots \otimes \underline{I}_{(m_{k-1},m_{k-1})} \otimes \underline{E}_{(m_k,m_k)}$$

$$\otimes \underline{I}_{(m_{k+1},m_{k+1})} \otimes \ldots \otimes \underline{I}_{(m_n x m_n)}] . \qquad (2.4d)$$

For any j and k ,

$$\int_0^{\kappa_j}\int_0^{\kappa_k}\underline{\Theta}(\kappa_1,\kappa_2,\ldots,\kappa_n)d\kappa_k d\kappa_j = \underline{E}_{\kappa_j\kappa_k}\underline{\Theta}(\kappa_1,\kappa_2,\ldots,\kappa_n)$$

$$= \underline{E}_{\kappa_k\kappa_j}\underline{\Theta}(\kappa_1,\kappa_2,\ldots,\kappa_n)$$

$$= \underline{E}_{\kappa_j}\underline{E}_{\kappa_k}\underline{\Theta}(\kappa_1,\kappa_2,\ldots,\kappa_n) \qquad (2.4.e)$$

$$= \underline{E}_{\kappa_k}\underline{E}_{\kappa_j}\underline{\Theta}(\kappa_1,\kappa_2,\ldots,\kappa_n),$$

where

$$\underline{E}_{\kappa_k}\underline{E}_{\kappa_j} = \underline{E}_{\kappa_j}\underline{E}_{\kappa_k} = [\underline{I}_{(m_1 x m_1)} \otimes \underline{I}_{(m_2 x m_2)} \otimes \ldots \otimes \underline{E}_{(m_j x m_j)} \otimes \underline{E}_{(m_k x m_k)}$$

$$\otimes \ldots \underline{I}_{(m_n x m_n)}] .$$

Finally,

$$\int_0^{\kappa_1}\int_0^{\kappa_2}\cdots\int_0^{\kappa_n} \underline{\Theta}(\kappa_1,\kappa_2,\ldots\kappa_n)\,d\kappa_1 d\kappa_2\ldots d\kappa_n$$

$$= \underline{E}_{\kappa_1\kappa_2\ldots\kappa_n}\underline{\Theta}(\kappa_1,\kappa_2,\ldots,\kappa_n)$$

$$= \underline{E}_{\kappa_1}\underline{E}_{\kappa_2}\cdots\underline{E}_{\kappa_n}\,\underline{\Theta}(\kappa_1,\kappa_2,\ldots\kappa_n)$$

$$= \left[\underline{E}_{(m_1 x m_1)} \otimes \underline{E}_{(m_2 x m_2)} \otimes \cdots \otimes \underline{E}_{(m_n x m_n)}\right] \underline{\Theta}(\kappa_1,\kappa_2,\ldots,\kappa_n). \tag{2.4f}$$

2.1.8. Operational matrices for fractional calculus via PCBF

The operational matrices for integration and differentiation, $\underline{E}$ and $\underline{E}^{-1}$ respectively in the domain of $\theta_i(t)$, $i=1,2,\ldots,m$, may be used to develop their fractional powers viz., $(\underline{E})^{1/l}$ and $(\underline{E})^{-1/l}$ respectively. Then, any fractional power may be realised to obtain operational matrices of the form $\underline{E}^{j/l}$ and $\underline{E}^{-j/l}$ respectively by repeated multiplication of $\underline{E}^{1/l}$ and $\underline{E}^{-1/l}$ respectively. Here i and l are integers.

The fractional power of the matrices $\underline{E}$ and $\underline{E}^{-1}$ to the degree 1/l each will be realised through the application of the Cayley Hamilton Theorem: i.e., every matrix satisfies its characteristic equation. Irrespective of the kind of PCBF, the eigenvalues of $\underline{E}$, $\underline{E}_\beta$, $\underline{E}_w$ and $\underline{E}_h$ are all the same. The eigenvalue is $\frac{1}{2m}$ repeated m times. Similarly, the eigen values of $\underline{E}^{-1}$, $\underline{E}_\beta^{-1}$, $\underline{E}_w^{-1}$ and $\underline{E}_h^{-1}$ are all the same. In each case, the value is 2m repeated m times. This is because these matrices are all related to one another by similarity transformations which keep the eigenvalues unaltered. A detailed discussion of these aspects will be presented with BPF as the basis. All the results may be inferred in other cases such as WF and HF through the similarity transformations such as those in section 1.10. The subscripts β, w and h in the above mentioned matrices signify that the operational matrix is with respect to BPF, WF and HF respectively.

2.2. Errors due to approximate operational matrices |W 53|

Consider a function f(t) on $[0,1)$ represented in terms of an m-set of PCBF. In this approximate representation let the constant (average) value of f(t) in the subinterval $\left[\frac{i-1}{m}, \frac{i}{m}\right]$ be $f_{\beta i}$.

2.2.1. Error in representation of the integral of a set of PCBF

The integral $\int_0^t f_{\beta i}dt$, $t\epsilon\left[\frac{i-1}{m}, \frac{i}{m}\right]$, is a ramp on the subinterval. The corresponding re-representation has an average value of $f_{\beta i}/2m$. The error in approximating the ramp by this constant value over the subinterval

$$\mu_i(t) = \frac{f_{\beta i}}{2m} - f_{\;i}(t-\frac{i-1}{m}),\quad t\epsilon\left[\frac{i-1}{m}, \frac{i}{m}\right) .$$

The norm of this error on the entire interval

$$\| \mu(t) \| \leq \frac{1}{2\sqrt{3}m} \sup t_z(f) \qquad . \tag{2.5}$$

The situation is shown in Fig. 2.1.

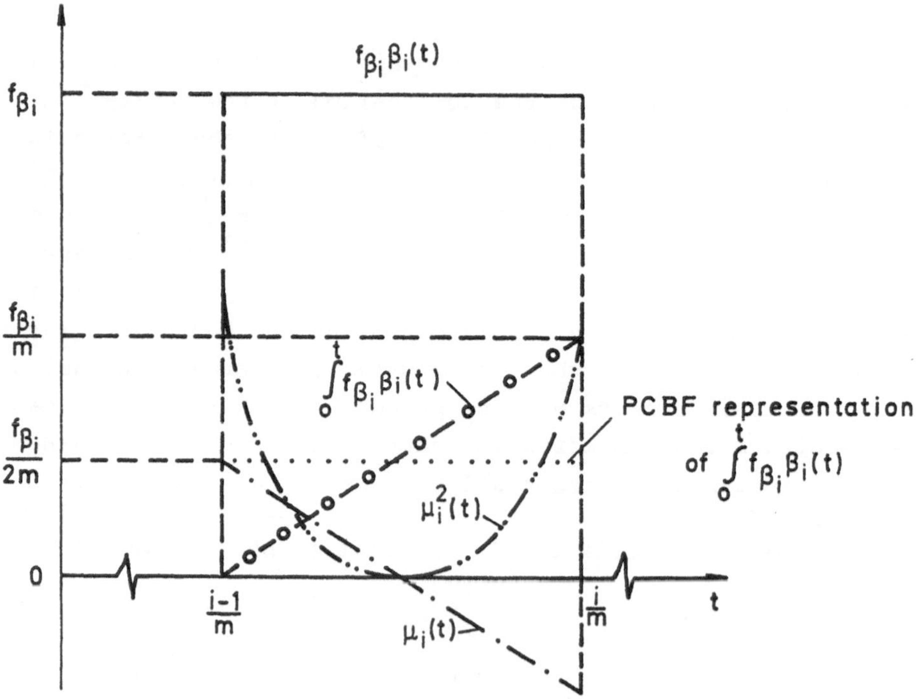

Fig. 2.1. Integration of PCBF and error in rerepresentation in terms of original PCBF

2.2.2. Error due to the delay operational matrix

The error

$$d(t) = \underline{f}_{\beta}^{T}\ \underline{\Theta}(t-\tau) - \underline{f}_{\beta}^{T}\underline{D}\ \underline{\Theta}(t)$$

is such that

$$||d||^2 \leq \frac{\alpha(1-\alpha)}{m}\ [\sup_{Z_\tau}(f) - \operatorname{Inf}_{Z_\tau}(f)]^2, \qquad (2.6)$$

where

$$\tau = \frac{N+\alpha}{m}\ ,\ 0<\alpha<1,\ N=0,1,2,\ldots,m-1,$$

and Z_τ is the interval $[-\tau,1)$. $||d||$ vanishes if $\alpha=0$ or 1 and has a maximum value at $\alpha=1/2$. It is, therefore, desirable to choose m such that for the given τ, α is either zero or unity. For further details see [W53, W66]. Fig. 2.2. shows the situation graphically.

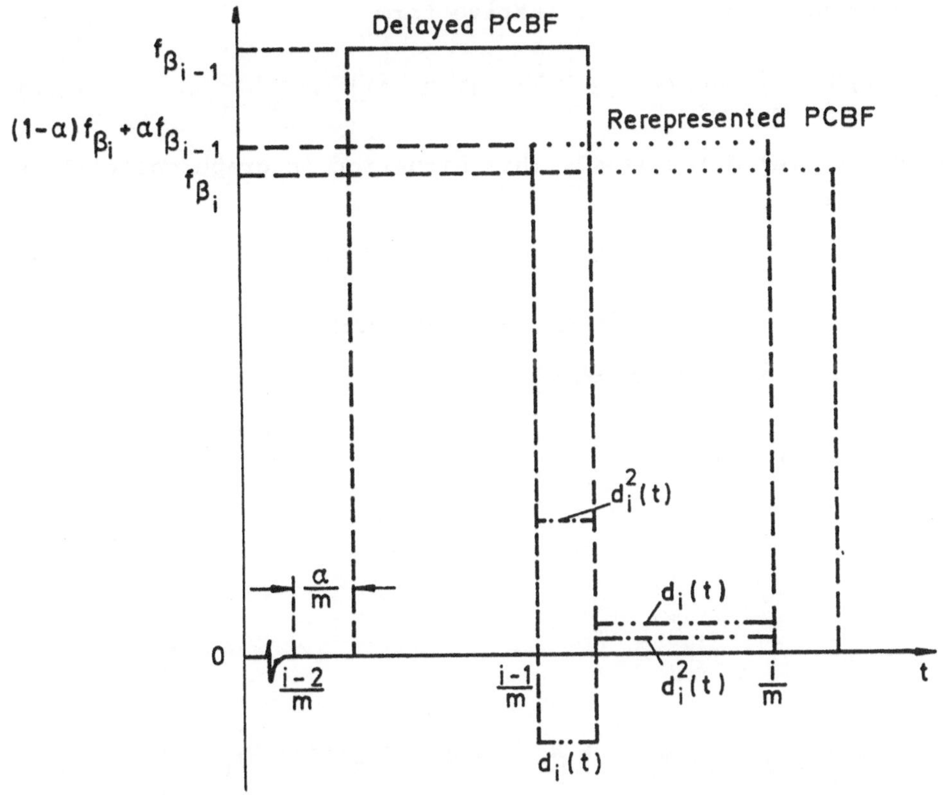

Fig. 2.2. Rerepresentation error in the i-th segment of a delayed PCBF

2.2.3. Error due to the stretch operational matrix

When $\underline{\Theta}(t/\lambda)$ is expressed in terms of $\underline{\Theta}(t)$, the resultant error is actually due to a uniformly increasing delay. Let the unit interval consist of ρ blocks each with Ω segments such that $m=\rho\Omega$. Let the i-th segment be in the first block. Then it can be shown [W66] that the error in this segment, $e_{si}(t)$, is such that

$$\|e_{si}(t)\|^2 = \frac{(f_{\beta_i}-f_{\beta_{i-1}})^2}{m}[1-(i-1)(\lambda-1)](i-1)(\lambda-1)$$

$$= \frac{\dot{f}^2(\frac{i-1}{m})}{m^3}[(i-1)(\lambda-1)-(i-1)^2(\lambda-1)^2] \quad .$$

This error taken over the interval t_z, $e_s(t)$, is such that

$$\|e_s(t)\| \leq \frac{\sup(\dot{f})}{\sqrt{2}\,m}(\tfrac{2}{3} - \tfrac{1}{3}\lambda)^{1/2} \quad . \tag{2.7}$$

We can express this in an alternative form

$$\|e_s(t)\| \leq \frac{1}{m\sqrt{6}} \sup_{t_z}(\dot{f}) \; [2^{-1/2}+(\lambda-1)(2\Omega-1)] \quad . \tag{2.8}$$

The above are valid for $1<\lambda<2$. This situation is graphically illustrated in Fig. 2.3.

2.3. Operations and operational matrices with BPF

2.3.1. Addition (or Subtraction)

If

$$e(t) = f(t) \pm g(t) \;,$$

then,

$$\underline{c}_\beta = \underline{f}_\beta \pm \underline{g}_\beta \;,$$

where,

$$\underline{f}_\beta = [f_{\beta_1}, f_{\beta_2}, \ldots, f_{\beta_m}]^T \;, \; \underline{g}_\beta = [g_{\beta_1}, g_{\beta_2}, \ldots, g_{\beta_m}]^T \;,$$

and

$$\underline{c}_\beta = [c_{\beta_1}, c_{\beta_2}, \ldots, c_{\beta_m}]^T .$$

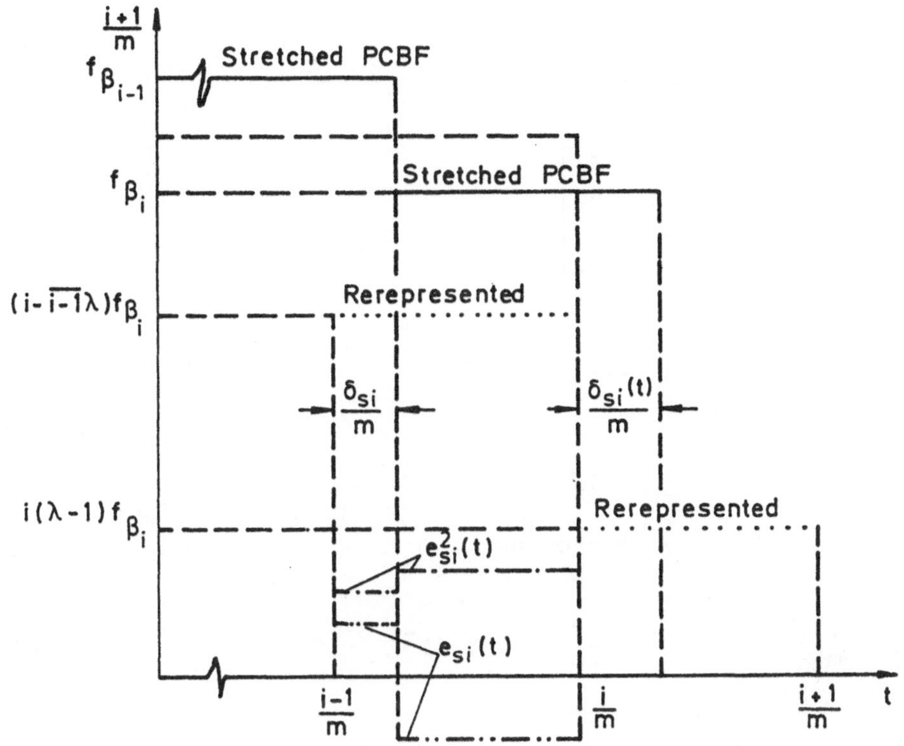

Fig. 2.3. Error in rerepresentation of stretched PCBF in terms of the original PCBF

2.3.2. Mutliplication (or division)

If

$$c(t) = f(t) \overset{\times}{\div} g(t) ,$$

then ,

$$c_{\beta_i} = f_{\beta_i} \overset{\times}{\div} g_{\beta_i} , \quad i=1,2,\ldots,m .$$

2.3.3. Integration with respect to 't'

Consider $\int_0^t \underline{\beta}(t)dt$ with m=4. Integrating each $\beta_i(t)$ in the integrand, we get functions as shown in Fig. 2.4. Expressing these in terms of the original BPF, we get

$$\int_0^t \underline{\beta}(t)dt = \underline{E}_\beta\, \underline{\beta}(t), \text{ where}$$

$$\underline{E}_\beta = \frac{1}{4}\begin{bmatrix} 1/2 & 1 & 1 & 1 \\ 0 & 1/2 & 1 & 1 \\ 0 & 0 & 1/2 & 1 \\ 0 & 0 & 0 & 1/2 \end{bmatrix}.$$

In general,

$$\underline{E}_\beta = \frac{1}{m}\begin{bmatrix} 1/2 & 1 & 1 & \dots & 1 \\ & 1/2 & 1 & \dots & 1 \\ & & & \dots & \\ & 0 & & & 1/2 \end{bmatrix}, \tag{2.9}$$

which may be concisely written as

$$\underline{E}_\beta = \frac{1}{m}\left[\frac{\underline{I}}{2} + \sum_{i=1}^{m-1} \underline{\Delta}^i\right] = \frac{1}{2m}(\underline{I}+\underline{\Delta})(\underline{I}-\underline{\Delta})^{-1}, \tag{2.10}$$

where

$$\underline{\Delta}_{(m\times m)} = \left[\begin{array}{c|c} 0 & \underline{I}_{(m-1)\times(m-1)} \\ \hline 0 & 0 \end{array}\right],$$

$$\underline{\Delta}^i = 0,\ i \geq m.$$

2.3.4. Differentiation with respect to 't'

The BPF spectrum of df/dt may be written as a special case of (2.2) as

$$\underline{f}_{d\beta} = \underline{E}_\beta^{-1}(\underline{f}_\beta - f(0)\underline{u}_\beta), \tag{2.11}$$

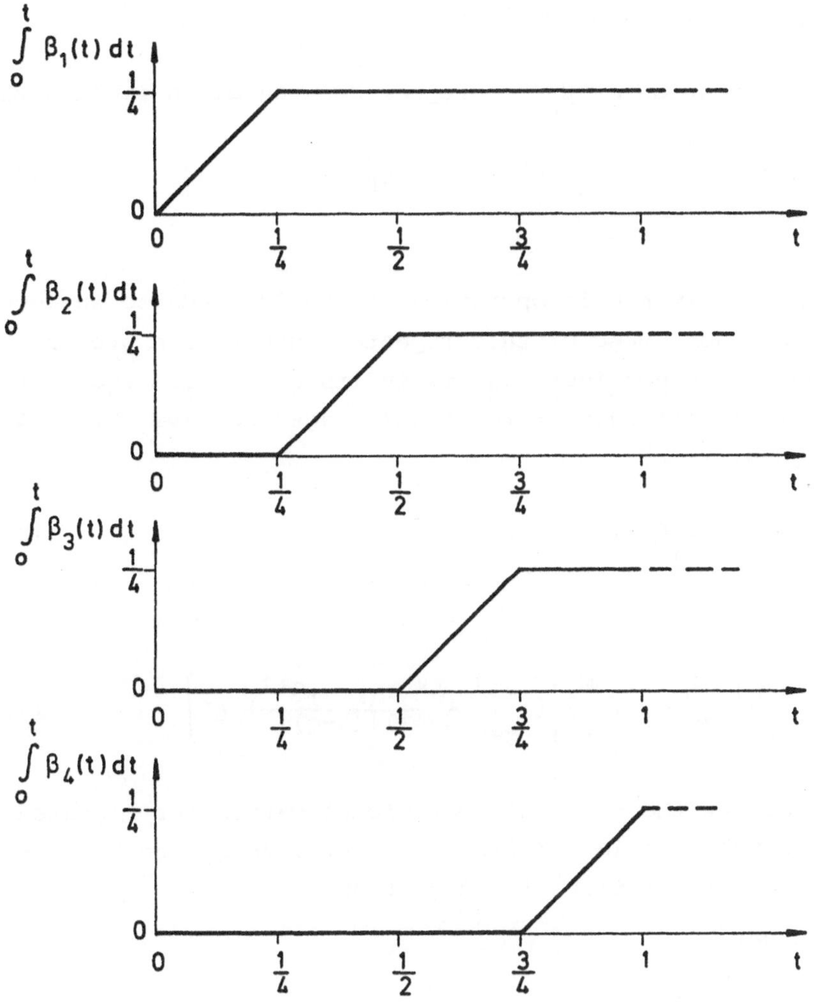

Fig. 2.4. Integrals of a set of 4 block pulse functions

where $\underline{u}_\beta$ is the BPF spectrum of a unit step function, and

$$\underline{E}_\beta^{-1} = 2m(\underline{I}-\underline{\Delta})(\underline{I}+\underline{\Delta})^{-1}$$

$$= 2m\ (\underline{I}-2\underline{\Delta}+2\underline{\Delta}^2+\ldots)\quad . \tag{2.12}$$

2.3.5. Repeated integration [W44]

Repeated application of $\underline{E}_\beta$ for repeated integration implies that

$$\underline{\beta}_{(j)}(t) \triangleq \underbrace{\int_0^t \int_0^t \dots \int_0^t}_{j\text{-times}} \underline{\beta}(t)dt^j \simeq \underline{E}_\beta^j \, \underline{\beta}(t) \quad . \tag{2.13}$$

This algebraically simple operation naturally leads to an accumulation of errors at each stage of this repeated integration process. On the other hand, if we consider $\underline{\beta}_{(j)}(t)$ in its exact form and re-represent it in terms of $\underline{\beta}(t)$ only at the final stage, we have the result in the form [W44]

$$\underline{\beta}_{(j)}(t) \simeq \underline{E}_{\beta_j} \underline{\beta}(t), \tag{2.14}$$

in which

$$\underline{E}_{\beta_j} = \frac{1}{m^j}\left[\frac{\underline{I}}{(j+1)!} + \sum_{r=1}^{m-1}\left\{\sum_{q=o}^{j-1} \frac{r^{q+1}-(r-1)^{q+1}}{(q+1)!(j-q)!}\right\} \underline{\Delta}^r\right] \quad , \tag{2.15}$$

and is termed as the one shot operational matrix for repeated integration (OSOMRI). Table 2.1 lists the matrices $\underline{E}_\beta^j$ and $\underline{E}_{\beta_j}$ for certain frequencies of repetition of integration.

We notice that

$$\underline{E}_\beta^2 - \underline{E}_{\beta_2} = \frac{1}{2m^2}\underline{I} \quad ,$$

$$\underline{E}_\beta^3 - \underline{E}_{\beta_3} = \frac{2}{3m^3}\underline{E}_\beta \quad ,$$

$$\underline{E}_\beta^4 - \underline{E}_{\beta_4} = \frac{1}{m^4}(4\underline{E}_\beta^2 - \frac{\underline{I}}{60}) = \frac{1}{m^4}(4\underline{E}_{\beta_2} + \frac{\underline{I}}{80}) \text{ and so on.}$$

j	$\underline{E}_{\beta}^{j}$	$\underline{E}_{\beta_j}$
1.	$\frac{1}{m}\left[\frac{\underline{I}}{2} + \sum_{r=1}^{m-1} \underline{\Delta}^r\right]$	$\frac{1}{m}\left[\frac{\underline{I}}{2} + \sum_{r=1}^{m-1} \underline{\Delta}^r\right]$
2.	$\frac{1}{m^2}\left[\frac{\underline{I}}{4} + \sum_{r=1}^{m-1} r\underline{\Delta}^r\right]$	$\frac{1}{m^2}\left[\frac{\underline{I}}{3!} + \sum_{r=1}^{m-1} r\underline{\Delta}^r\right]$
3. (For m=4, say)	$\frac{1}{64}\left[\frac{\underline{I}}{8}+\frac{3}{3}\underline{\Delta}+\frac{9}{4}\underline{\Delta}^2+\frac{19}{4}\underline{\Delta}^3\right]$	$\frac{1}{m^3}\left[\frac{\underline{I}}{4!}+\sum_{r=1}^{m-1}\left(\frac{1+6r^2}{12}\right)\underline{\Delta}^r\right]$
4. (For m=4, say)	$\frac{1}{256}\left[\frac{\underline{I}}{16}\;\frac{1}{2}\;+2\underline{\Delta}^2+\frac{11}{2}\underline{\Delta}^3\right]$	$\frac{1}{m^4}\left[\frac{\underline{I}}{5!}+\sum_{r=1}^{m-1}\left(\frac{2r^3+r}{12}\right)\underline{\Delta}^r\right]$

Table 2.1. Matrices $\underline{E}_{\beta}^{j}$ and $\underline{E}_{\beta_j}$

The errors resulting from the application of $\underline{E}_{\beta}^{j}$ and $\underline{E}_{\beta_j}$ are respectively

$$\mu_r(t) = \left(\frac{t^j}{j!} - \frac{1}{2^j m^j}\right), \text{ and}$$

$$\mu_o(t) = \left(\frac{t^j}{j!} - \frac{1}{(j+1)!m^j}\right).$$

It is possible to show that [W44]

$$\varepsilon_j \triangleq \frac{\|\mu_r\|}{\|\mu_o\|} = \frac{(j+1)^{1/2}}{j\ 2^j}\left[(j+1)2^{2j}+j!(2j+1)\{(j+1)!-2^{j+1}\}\right]^{1/2}$$

$$\geq 1 \quad \text{for } j=1,2,3,... \tag{2.16}$$

Table 2.2. shows ε_j for some values of j indicating the superiority of OSOMRI over E_β^j.

j	ε_j
1	1.0000000
2	1.1456439
3	2.0275875
4	4.9765073
5	14.2881060

Table 2.2. The norm ratio ε_j

2.3.6. Delay operational matrices

Given an m-vector $\underline{\beta}(t)$, $\underline{\beta}(t-\tau)$ for constant τ may be expressed in terms of $\underline{\beta}(t)$ through a linear and sometimes singular transformation. For the sake of illustration for m=4, Fig. 2.5 shows the situation. We note that

$$\underline{\beta}(t-\tau) = \underline{D}_\beta \underline{\beta}(t),$$

where $\underline{D}_\beta$ is the BPF delay matrix given by

$$\underline{D}_\beta = (1-\alpha)\underline{\Delta}^N + \alpha\underline{\Delta}^{N+1} . \tag{2.17}$$

Here $\tau = \frac{N+\alpha}{m}$, $0\leq\alpha\leq 1$, N any positive integer.

It should be recalled here that α should be adjusted to be zero (or 1) for the error in the delay operation to vanish. This may be done by suitable time scaling and segmentation for the given τ.

When τ is negative, it means to advance a signal by a known time period, as for instance, in the case of computation of correlation functions. The related matrix for 'advance' happens to be the transpose of $\underline{D}_\beta$. Thus, the operational matrix for advance,

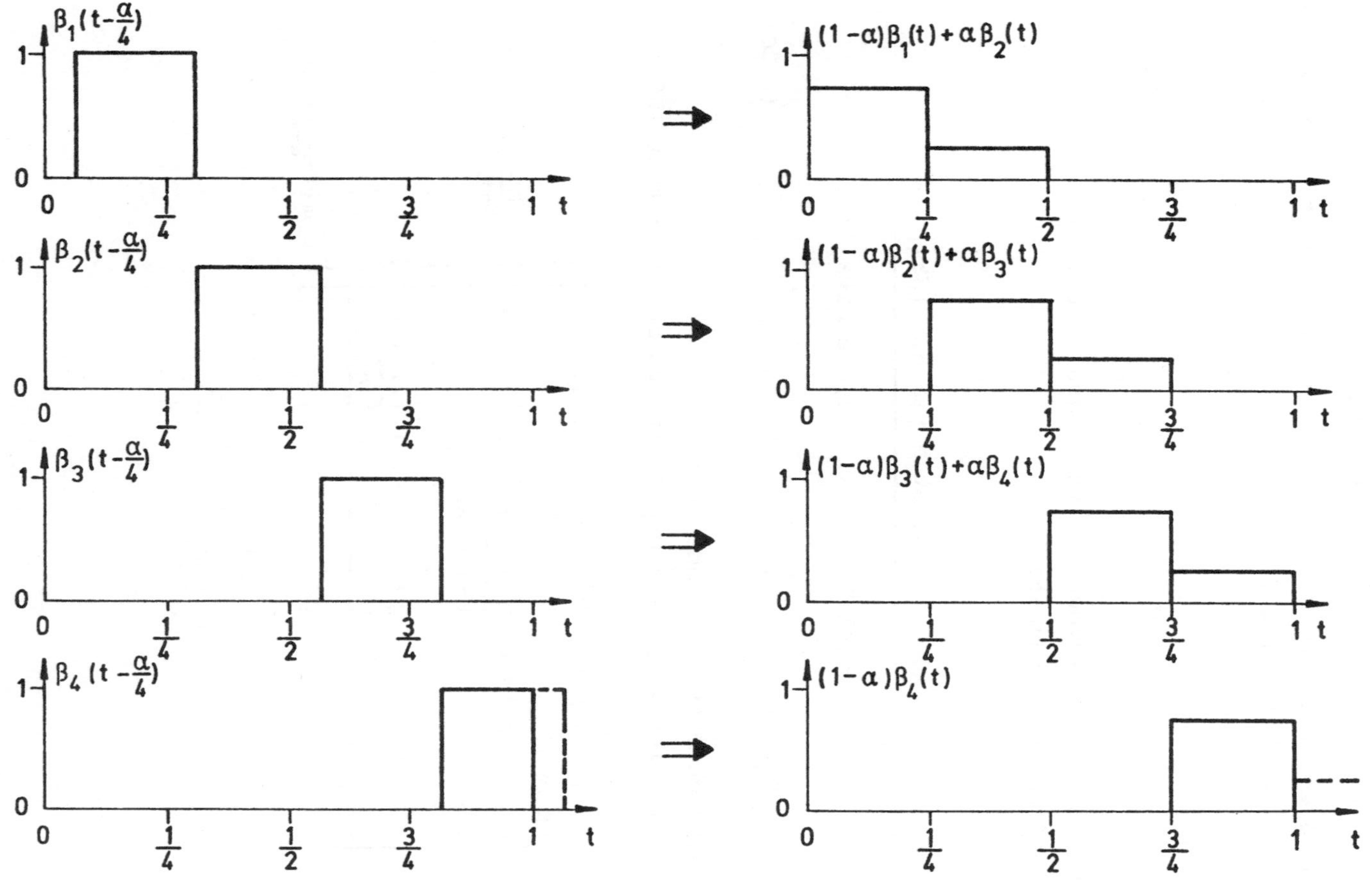

Fig. 2.5. A 4-set of BPF delayed by $\frac{\alpha}{m}$, $\alpha=\frac{1}{4}$ and its re-representation in terms of the original set of BPF

$$\underline{D}_\beta^- = [\underline{D}_\beta]^T \, . \tag{2.18}$$

When $\alpha=0$, $\underline{D}_\beta = \underline{\Delta}^N$, and $\underline{D}_\beta^- = \underline{\nabla}^N$, where $\underline{\nabla} = \underline{\Delta}^T$.

The matrices $\underline{\Delta}$ and $\underline{\nabla}$ follow the laws of algebraic powers only in one direction, i.e., either $N=1,2,\ldots$ or $N=-1,-2,\ldots$ It is important to note that

$$\underline{D}_\beta^- \, \underline{D}_\beta \neq \underline{I} \, . \tag{2.19}$$

This is because certain segment of the signal overflows the interval t_z as a result of delay or advance.

2.3.7. Operational matrices for stretch via BPF [W54]

If the variable t is stretched to t/λ, $\lambda>1$, we have

$$\underline{\beta}(t/\lambda) = \underline{S}_\beta \, \underline{\beta}(t), \tag{2.20}$$

where the stretch matrix depends on λ.

If $\lambda=N+\kappa$, $N\geq 1$ (an integer), and $0\leq\kappa\leq 1$, it can be shown that [W54]

$$\underline{S}_\beta = \left[\begin{array}{cccccc} \underbrace{1 \;\; 1\ldots1}_{N \text{ ones}} & \kappa & 0 \; 0 \; \ldots\ldots\ldots\ldots & & & 0 \\ 0 \;\; 0\ldots0 & 1-\kappa & \underbrace{1 \;\; 1\ldots1}_{N-1 \text{ ones}} & 2\kappa & & 0 \\ \ldots & \ldots & \ldots & \ldots & & \\ & & |1-(k-1)\kappa| & \underbrace{1 \;\; 1\ldots1}_{N-1 \text{ ones}} & k\kappa & \\ \hline & & 0 & & & \end{array}\right] \begin{array}{l} \left.\vphantom{\begin{array}{c}1\\1\\1\\1\end{array}}\right\} \frac{m}{N+\kappa} \text{ rows} \\ \left.\right\} m\left(1-\frac{1}{N+\kappa}\right) \text{ rows} \end{array} \tag{2.21}$$

As a special case, if $\lambda = 1+\kappa \leq \frac{m}{m-1}$

$$\underline{S}_\beta = \begin{bmatrix} 1 & \lambda-1 & 0 & 0 \ldots\ldots\ldots\ldots & 0 \\ 0 & 2-\lambda & 2(\lambda-1) & 0 \ldots\ldots\ldots\ldots & 0 \\ 0 & 0 & 0 & (k-\overline{k-1}\lambda) \quad k(\lambda-1)\ldots & 0 \\ \ldots & \ldots & \ldots & \ldots & \ldots \\ & & 0 & & (m-\overline{m-1}\lambda) \end{bmatrix} \quad . \qquad (2.22)$$

If $\lambda = \frac{m}{m-1}$, the last row of the above matrix will contain all zeros. In general, if $\lambda = \frac{m}{m-\sigma}$, the last σ rows contain zeros as their elements. This is so because of stretching. Fig. 2.6 shows the case of 5 BPF stretched by λ=1.25.

2.3.8. Operational matrices for fractional calculus via BPF and inversion of irrational Laplace transforms [W16]

Let us recall (2.10) which says that

$$\underline{E}_\beta = \mathcal{E}_\beta(\Delta) = \frac{1}{2m}(\underline{I}+\underline{\Delta})(\underline{I}-\underline{\Delta})^{-1} \quad .$$

If δ is an eigenvalue of $\underline{\Delta}$, the eigenvalue of $\underline{E}_\beta$ is given by

$$\mathcal{E}_\beta(\delta) = \frac{1}{2m}\frac{(1+\delta)}{(1-\delta)} \quad . \qquad (2.22a)$$

Since all the eigenvalues of $\underline{\Delta}$ are zero, the eigenvalues of $\underline{E}_\beta$, viz.,

$$e_\beta = \mathcal{E}_\beta(0) = 1/2m.$$

<u>Fractional differentiation and integration</u>

The eigenvalue of $\underline{E}_\beta^{-1}$, e_β^{-1}, is expressed as

$$\bar{e}_\beta = 2m\,\frac{(1-\delta)}{(1+\delta)} \quad . \qquad (2.22b)$$

In view of (2.22b),

$$(\bar{e}_\beta)^{1/l} = \left(m\,\frac{1-\delta}{1-\delta}\right)^{1/l} \quad . \qquad (2.22c)$$

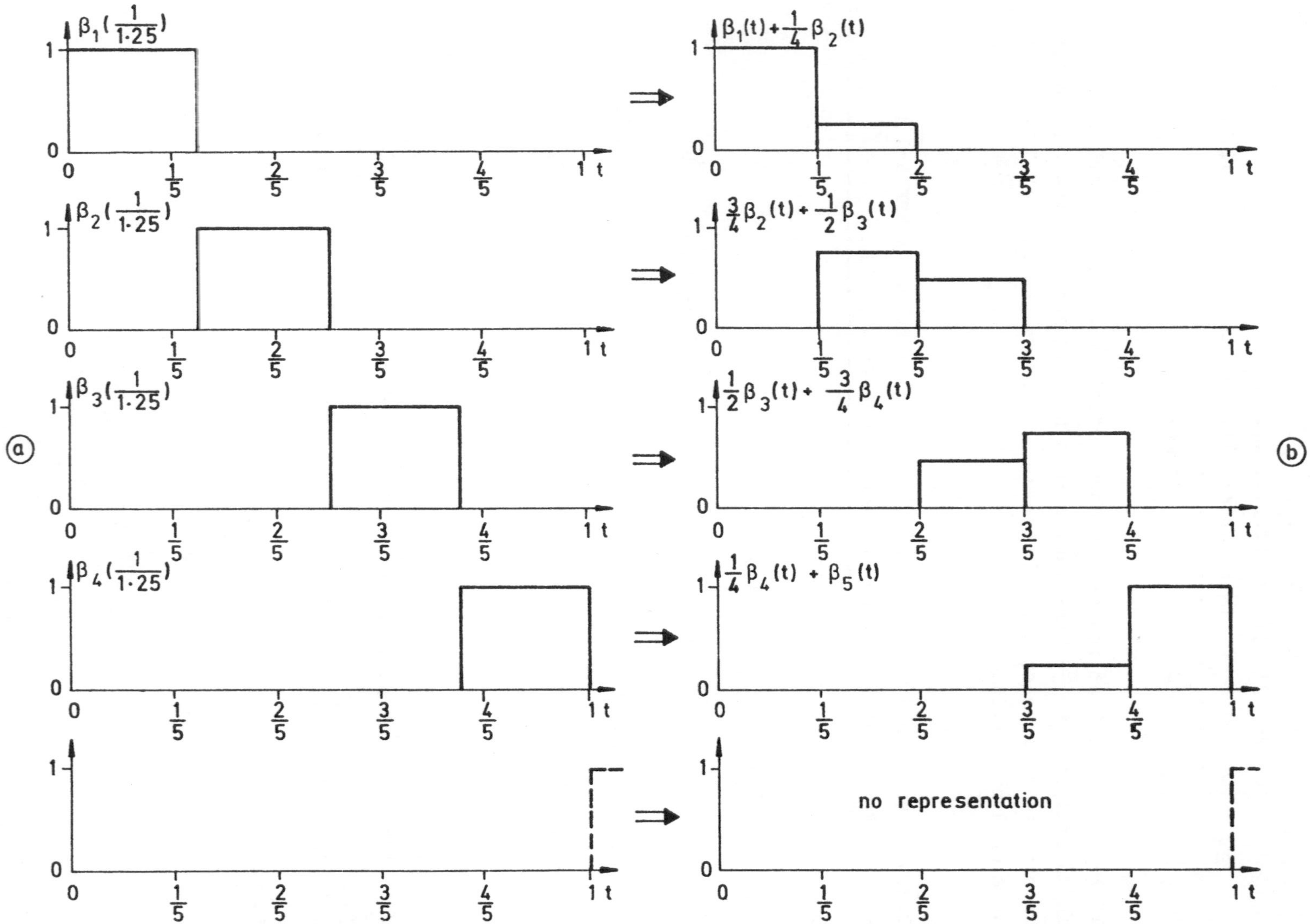

Fig. 2.6. A 5-set of BPF stretched by λ=1.25 and its rerepresentation in terms of the original set

The RHS of (2.22c) may be expanded as a polynominal of degree (m-1) in δ,i.e.,

$$(\bar{e}_\beta)^{1/1} = (2m)^{1/1}\, \bar{\mathcal{E}}_{1,m}(\delta) \; . \tag{2.22d}$$

Equation (2.22d) implies that the operational matrix

$$(E_\beta)^{-1/1} = (2m)^{1/1}\; \bar{\mathcal{E}}(\Delta). \tag{2.22e}$$

As an example,consider the operational matrix corresponding to $\sqrt{s}$ with m=4. In this case,

$$\bar{\mathcal{E}}_{1,m} = \bar{\mathcal{E}}_{2,4}(\delta) = \left(\frac{1-\delta}{1+\delta}\right)^{1/2} = 1-\delta+\frac{1}{2}\delta^2-\frac{1}{2}\delta^3 \; .$$

This gives

$$(\underline{E}_\beta)^{-1/2} = \sqrt{8}\,(\underline{I}-\underline{\Delta}+\frac{1}{2}\underline{\Delta}^2-\frac{1}{2}\underline{\Delta}^3) \; . \tag{2.22f}$$

In the case of fractional integration,

$$(e_\beta)^{1/1} = \left|\frac{(1-\delta)}{2m(1-\delta)}\right|^{1/1} = \frac{1}{(2m)^{1/1}}\, \bar{\bar{\mathcal{E}}}_{1,m}\;(\delta) \; . \tag{2.22g}$$

For instance,if we wish the equivalent of $1/\sqrt{s}$ with m=4,

$$\bar{\bar{\mathcal{E}}}_{2,4}(\delta) = 1+\delta+\frac{1}{2}\delta^2+\frac{1}{2}\delta^3 \; .$$

$$\therefore\; (\underline{E}_\beta)^{1/2} = \frac{1}{\sqrt{8}}(\underline{I}+\underline{\Delta}+\frac{1}{2}\underline{\Delta}^2+\frac{1}{2}\underline{\Delta}^3). \tag{2.22h}$$

<u>Inversion of irrational Laplace transforms through operational matrices for fractional calculus [W16]</u>

A Laplace transform function F(s) is first written as $F_1(s)\frac{1}{s}$. The irrational function F(s) is time scaled, if necessary, for convenience. $\bar{e}_\beta(\delta)$ is then substituted for s in the expression $F_1(s)$. The result is then expanded in power series of δ.In the next step, 1 and δ are replaced by $\underline{I}$ and $\underline{\Delta}$ to give the required operational matrix. This matrix is then operated upon the unit step function in BPF, i.e., a vector $[1\ 1\ \ldots\ 1]^T\ \underline{\beta}(t)$ to give the numerical inversion of F(s).

For example, let $F(s) = \frac{1}{s}\exp(-\frac{1}{s+1})$, m=4, and $0 \le t \le 8$.

$$F_1(\delta) = \exp\left[-(\tfrac{1-\delta}{1+\delta})/\left[(\tfrac{1-\delta}{1+\delta})+1\right]\right]$$

$$= e^{-1/2}\left[1 + \frac{\delta}{2} + \frac{(\delta/2)^2}{2!} + \frac{(\delta/2)^3}{3!} + \ldots\right]$$

$$= e^{-1/2}\left[1 + 0.5\delta + 0.125\delta^2 + 0.0208\delta^3\right]$$

$$F_1(\underline{\Delta}) = e^{-1/2}\left[\underline{I} + 0.5\underline{\Delta} + 0.125\underline{\Delta}^2 + 0.0208\underline{\Delta}^3\right]$$

$$= 0.6065\underline{I} + 0.3033\underline{\Delta} + 0.07852\underline{\Delta}^2 + 0.01264\underline{\Delta}^3 \ .$$

The inverse Laplace transform of F(s) is then

$$\mathcal{L}^{-1}\left[F_1(s)\ \frac{1}{s}\right] = (1\quad 1\quad 1\quad 1)\ \left[F_1(\underline{\Delta})\right]$$

$$= [0.6065,\ 0.9098\ ,\ 0.9856,\ 0.9982]\ \underline{\beta}(t) \ .$$

A similar treatment via WF directly follows from the similarity transformations. The results will be identical. The reader may try the above with WF as an exercise. Chen et al [W16] discussed such matrices with WF as the basis.

2.4. Operations and operational matrices with Walsh functions

2.4.1. Addition (or subtraction)

If $c(t) = f(t) \pm g(t)$,

then, $\underline{c}_W = \underline{f}_W + \underline{g}_W$

where $\underline{f}_W$, $\underline{g}_W$ and $\underline{c}_W$ are m-vectors of Walsh spectral components of f(t), g(t) and c(t) respectively.

2.4.2. Multiplication

If $c(t) = f(t)\ g(t)$, and if $f(t) = \underline{f}_W^T\ \underline{w}(t)$ and $g(t) = \underline{g}_W^T\ \underline{w}(t)$, we are to find $\underline{c}_W$ such that $c(t) = \underline{c}_W^T\ \underline{w}(t)$. We note that the product of any two Walsh functions $w_i(t)$ and $w_j(t)$ is another Walsh function $w_l(t)$ such that l, when expressed in binary form, is equal to the sum of i and j. That is, $l=i\oplus j$ by modulo-2 addition. With this as the basis, we will obtain a general multiplication operator in terms of Walsh spectra as follows: We first write two matrices whose elements are values of indices say with m=8, i=0,1,2,...,7 of $w_i(t)$ as

$$\underline{i}_F = [0 \quad 1 \quad 2 \quad .... \quad 7]^T,$$

and

$$\underline{I}_G = \begin{bmatrix} 0 & 1 & 2 & 3 & 4 & 5 & 6 & 7 \\ 1 & 0 & 3 & 2 & 5 & 4 & 7 & 6 \\ 2 & 3 & 0 & 1 & 6 & 7 & 4 & 5 \\ 3 & 2 & 1 & 0 & 7 & 6 & 5 & 4 \\ 4 & 5 & 6 & 7 & 0 & 1 & 2 & 3 \\ 5 & 4 & 7 & 6 & 1 & 0 & 3 & 2 \\ 6 & 7 & 4 & 5 & 2 & 3 & 0 & 1 \\ 7 & 6 & 5 & 4 & 3 & 2 & 1 & 0 \end{bmatrix} .$$

Then,

$$i_C = \underline{I}_G\ \underline{i}_F \quad . \qquad (2.23)$$

Furthermore, a general recursive formula may be used to write $\underline{I}_G$ of any size from $I_{G(2x2)}$. To do this we begin with the basic (2x2) modules

$$\underline{I}_{G(2x2)} = \begin{bmatrix} 0 & 1 \\ 1 & 0 \end{bmatrix} , \text{ and}$$

$$\underline{I}_V = \begin{bmatrix} 1 & 1 \\ 1 & 1 \end{bmatrix} .$$

Then, for any positive integer k

$$
\underline{I}_{G(2^{k+1}x2^{k+1})} = \left[\begin{array}{c|c} \cdot\ \underline{I}_{G(2^kx2^k)} & 2^k\underline{I}_{V(2^kx2^k)} + \underline{I}_{G(2^kx2^k)} \\ \hline 2^k\underline{I}_{V(2^kx2^k)} + \underline{I}_{G(2^kx2^k)} & \underline{I}_{G(2^kx2^k)} \end{array}\right] . \quad (2.24)
$$

2.4.3. Integration w.r.t. t

With the first four components in $\underline{w}(t)$, $\int_0^t \underline{w}(t)dt$ is shown in Fig. 2.7 from which it is clear that

$$
\int_0^t \underline{w}(t)dt = \begin{bmatrix} 1/2 & -1/4 & -1/8 & 0 \\ 1/4 & 0 & 0 & -1/8 \\ 1/8 & 0 & 0 & 0 \\ 0 & 1/8 & 0 & 0 \end{bmatrix} \underline{w}(t) = \underline{E}_W\ \underline{w}(t) .
$$

In general,

$$
\underline{E}_{W(mxm)} = \left[\begin{array}{c|c} \underline{E}_{W(\frac{m}{2}x\frac{m}{2})} & -\frac{1}{2m}\underline{I}_{(\frac{m}{2}x\frac{m}{2})} \\ \hline \frac{1}{2}\underline{I}_{(\frac{m}{2}x\frac{m}{2})} & \underline{O}_{(\frac{m}{2}x\frac{m}{2})} \end{array}\right] = \frac{1}{m}\ \underline{T}_{BW}\ \underline{E}_\beta\ \underline{T}_{BW}^{-1} . \quad (2.25)
$$

The OSOMRI via WF

$$
\underline{E}_{Wj} = \frac{1}{m}\ \underline{T}_{BW}\ \underline{E}_{\beta j}\ \underline{T}_{BW}^{-1} . \quad (2.26)
$$

2.4.4. Differentiation

Given the Walsh spectrum of $f(t)$ as $\underline{f}_W$, then that corresponding to df/dt,

$$
\underline{f}_{dW} = \underline{E}_W^{-1}(\underline{f}_W - f(0)\ \underline{u}_W) , \quad (2.27)
$$

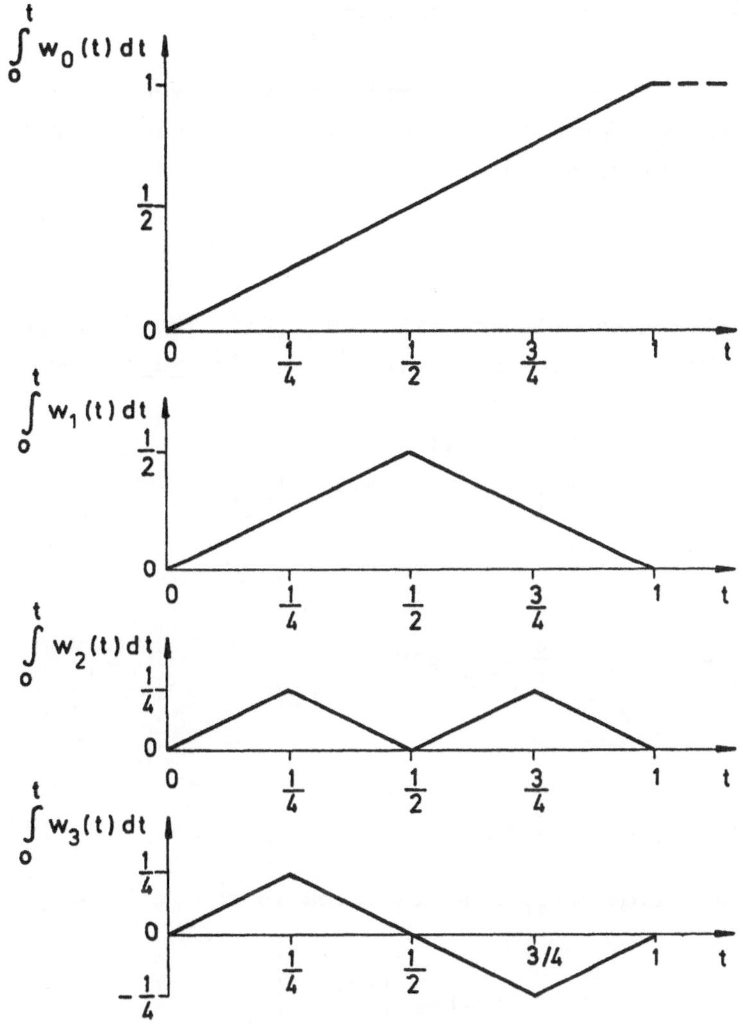

Fig. 2.7. Integrals of a set of 4 Walsh functions

where

$$\underline{u}_W = [1 \quad \underbrace{0 \ \ 0 \ \dots \ 0}_{m-1}]^T, \text{ and}$$

$$\underline{E}_W^{-1} = \frac{1}{m}\,\underline{T}_{BW}\,\underline{E}_\beta^{-1}\,\underline{T}_{BW}^{-1} \ . \tag{2.28}$$

2.4.5. Delay operational matrices [W36]

We focus our attention on the normal interval. We will, at the same time, ignore carry-over effects. These will be included in the final stage of computation. For a known constant τ, we write

$$\underline{w}(t-\tau) = \underline{D}_W \, \underline{w}(t) , \qquad (2.29a)$$

where $\underline{D}_W$ is the Walsh delay matrix (WDM) [W36]. The development of the WDM will be clear from Fig. 2.8. If $\tau = \frac{\alpha}{m}$, $0 \leq \alpha \leq 1$, then,

$$\underline{D}_{W(2x2)} = \underline{I}_{(2x2)} + \frac{\alpha}{2} \begin{bmatrix} -1 & -1 \\ 1 & -3 \end{bmatrix} = \underline{I}_{(2x2)} + \frac{\alpha}{2} \underline{\phi}_{d(2)} \text{ say.} \qquad (2.29b)$$

Similarly,

$$\underline{D}_{W(4x4)} = \underline{I}_{(4x4)} + \frac{\alpha}{4} \underline{\phi}_{d(4)} \text{ say.} \qquad (2.29c)$$

In general,

$$\underline{D}_{W(mxm)} = \underline{I}_{(mxm)} + \frac{\alpha}{m} \underline{\phi}_{d(m)} , \qquad (2.29d)$$

where the evaluation of $\underline{\phi}_{d(m)}$ is described in the following.
Let

$$\underline{\phi}_{d(2)}^{(-1)} = \begin{bmatrix} -1 & -1 \\ 1 & -3 \end{bmatrix}, \quad \underline{\phi}_{d(2)}^{(-3)} = \begin{bmatrix} -7 & 1 \\ -1 & -5 \end{bmatrix},$$

$$\underline{\phi}_{d(2)}^{(-5)} = \begin{bmatrix} -9 & -1 \\ 1 & -11 \end{bmatrix}, \quad \underline{\phi}_{d(2)}^{(-7)} = \begin{bmatrix} -15 & 1 \\ -1 & -13 \end{bmatrix} \quad \ldots$$

The $\underline{\phi}_{d(2)}$ matrix associated with any negative odd integer has its elements as follows:

$$\phi_{d\,12}^{(j-2)} = -1\phi_{d12}^{(j)} ; \quad \phi_{d\,21}^{(j-2)} = -1 \; \phi_{d\,21}^{(j)}$$

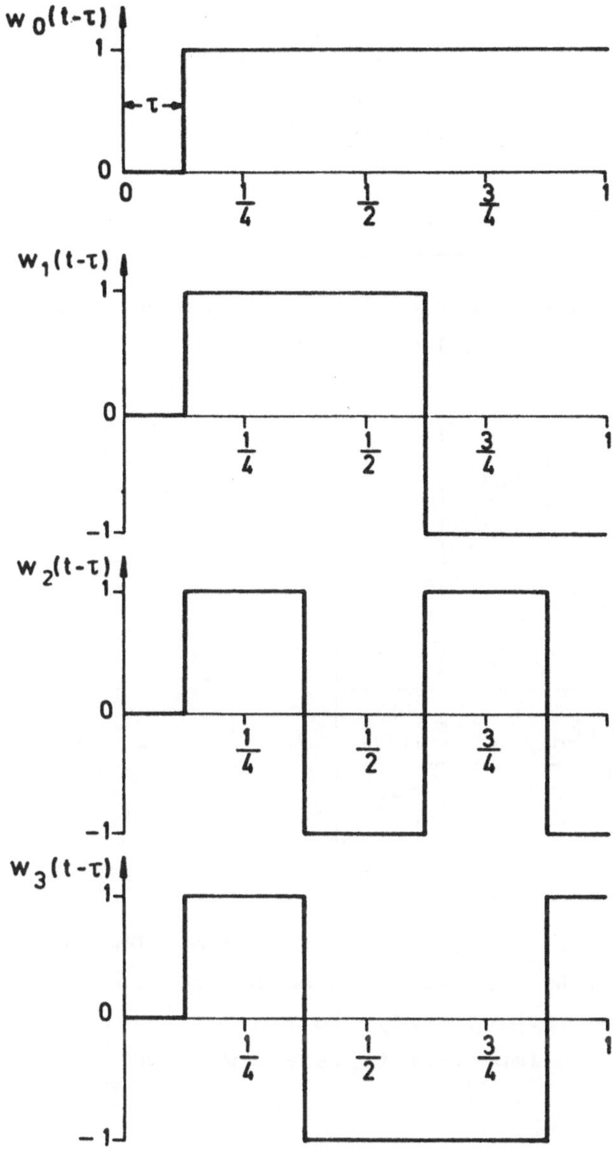

Fig. 2.8. Delayed Walsh functions ($m=4$, $\tau=\alpha/m$)

The diagonal elements take values from the sequence marked by the direction of arrows as below:

	j					
	1	3	5	7	9	....
ϕ_{d11}	-1	-7	-9	-15	-17	
	↓	↑	↓	↑	↓	
ϕ_{d22}	-3	-5	-11	-13	-19	

$\underline{\phi}_d$ matrices of higher dimension may be recursively built from those of lower dimension as follows.

$$\underline{\phi}_{d(4)} = \left[\begin{array}{c|c} \underline{\phi}_{d(2)}^{(-1)} & \underline{\phi}_{d(2)}^{(-1)} \\ \hline -\underline{\phi}_{d(2)}^{(-1)} & \underline{\phi}_{d(2)}^{(-3)} \end{array}\right] = \begin{bmatrix} -1 & -1 & -1 & -1 \\ 1 & -3 & 1 & -3 \\ 1 & 1 & -7 & 1 \\ -1 & 3 & -1 & -5 \end{bmatrix},$$

$$\underline{\phi}_{d(8)} = \left[\begin{array}{c|cc} \underline{\phi}_{d(4)} & \multicolumn{2}{c}{\underline{\phi}_{d(4)}} \\ \hline -\underline{\phi}_{(4)} & \begin{array}{c|c} \underline{\phi}_{d(2)}^{(-1)} & -\underline{\phi}_{d(2)}^{(-1)} \\ \hline \underline{\phi}_{d(2)}^{(-1)} & \underline{\phi}_{d(2)}^{(-5)} \end{array} \end{array}\right],$$

and so on.

If $\tau = \frac{N+\alpha}{m}$, $0 \leq N \leq m$, $0 \leq \alpha \leq 1$, the corresponding WDM's may be built in a similar manner. As has been pointed out before, it is desirable to adjust α to be zero or unity. In view of this, the structure of the WDM with $\alpha=0$ is important to us and will be discussed in the following.

WDM for $\tau=\frac{N}{m}$, N being an integer

By setting N=0, $\alpha=1$ in (2.29a and b), we get

$$\underline{D}^1_{W(2x2)} = \frac{1}{2}\begin{bmatrix} 1 & -1 \\ 1 & -1 \end{bmatrix}, \quad \underline{D}^1_{W(4x4)} = \frac{1}{4}\begin{bmatrix} 3 & -1 & -1 & -1 \\ 1 & 1 & 1 & -3 \\ 1 & 1 & -3 & 1 \\ -1 & 3 & -1 & -1 \end{bmatrix},$$

and so on. In general,

$$\underline{D}_{W\,(2^{k+1}x2^{k+1})}^{1} = \left[\begin{array}{c|c} \underline{D}_{W\,(2^k x2^k)}^{1} + \underline{D}_{W\,(2^k x2^k)}^{O} & \underline{D}_{W\,(2^k x2^k)}^{1} - \underline{D}_{W\,(2^k x2^k)}^{O} \\ \hline -(\underline{D}_{W\,(2^k x2^k)}^{1} - \underline{D}_{W\,(2^k x2^k)}^{O}) & -(\underline{D}_{W\,(2^k x2^k)}^{1} + \underline{D}_{W\,(2^k x2^k)}^{O}) \end{array}\right], \qquad (2.30)$$

where

$$\underline{D}_{W\,(2^k x2^k)}^{O} = \underline{I}_{(2^k x2^k)} .$$

$\underline{D}_W^N$ behaves as the N-th power of $\underline{D}_W^1$. That is ,

$$\underline{D}_{W\,(mxm)}^{N} = \left[\underline{D}_{W\,(mxm)}^{1}\right]^{N} . \qquad (2.31)$$

In the WDM, $\underline{D}_{W\,(mxm)}^{N}$, $O<N<m$,

the elements $d_{w_{ij}}$ are such that

$$\sum_{j=1}^{m} d_{w_{ij}} = O, \text{ for all } i$$

$$\sum_{i=1}^{m} d_{w_{ij}} = \underline{T}_{BW\,(N+1,j)}, \quad \text{for } i=1,2,\ldots,m$$

generating the (N+1)-th row of $\underline{T}_{BW}$.

$\underline{D}_{W\,(mxm)}^{1}$ is nilpotent with index m. That is

$$\underline{D}_{W\,(mxm)}^{N} = O, \; N\geq m .$$

Generation of $\underline{D}_{W\,(mxm)}^{N}$ through Kronecker product formula [W36]

$\underline{D}_{W\,(mxm)}^{N}$ may be generated from WDM of lower dimension at lower power through the Kronecker product formula given below, without resorting to matrix power operations:

$$\underline{D}_{W\,(2^{k+1}x2^{k+1})}^{2^k+M} = \underline{D}_{W\,(2^k x2^k)}^{M} \otimes \underline{D}_{W\,(2x2)}^{1} \quad . \tag{2.32}$$

For example,

$$\underline{D}_{W\,(8x8)}^{4} = \underline{D}_{W\,(4x4)}^{o} \otimes \underline{D}_{W\,(2x2)}^{1}$$

$$= \frac{1}{8}\begin{bmatrix} 4 & -4 & & & & & & \\ 4 & -4 & & & & & & \\ & & 4 & -4 & & & & \\ & & 4 & -4 & & & & \\ & & & & 4 & -4 & & \\ & & & & 4 & -4 & & \\ & & & & & & 4 & -4 \\ & & & & & & 4 & -4 \end{bmatrix} ,$$

$$\underline{D}_{W\,(8x8)}^{7} = \underline{D}_{W\,(4x4)}^{3} \otimes \underline{D}_{W\,(2x2)}^{1}$$

$$= \frac{1}{8}\begin{bmatrix} 1 & -1 & -1 & 1 & -1 & 1 & 1 & -1 \\ 1 & -1 & -1 & 1 & -1 & 1 & 1 & -1 \\ 1 & -1 & -1 & 1 & -1 & 1 & 1 & -1 \\ 1 & -1 & -1 & 1 & -1 & 1 & 1 & -1 \\ 1 & -1 & -1 & 1 & -1 & 1 & 1 & -1 \\ 1 & -1 & -1 & 1 & -1 & 1 & 1 & -1 \\ 1 & -1 & -1 & 1 & -1 & 1 & 1 & -1 \\ 1 & -1 & -1 & 1 & -1 & 1 & 1 & -1 \end{bmatrix} .$$

Generation of WDM for $\tau = 2N/m$ by Kronecker product formula [W36]

It is possible to show that

$$\underline{D}_{W\,(2^{k+1}x2^{k+1})}^{2N} = \underline{D}_{W\,(2x2)}^{o} \otimes \underline{D}_{W\,(2^k x2^k)}^{N} \quad . \tag{2.33}$$

This suggests that all even powers of $\underline{D}_{W\,(mxm)}^{1}$ are in block diagonal form.

2.4.6. Nonzero initial functions and carry-over considerations associated with WDM

The effect of delay on the single interval $[0,1)$ alone will not suffice if there is a nonzero initial function. The Walsh spectrum of the initial function on $[-\tau,0]$ should be properly carried over into $[0,1)$. Let g(t) be a function as shown in Fig. 2.9(a) subjected to a delay τ. The Walsh spectrum of $g(t-\tau)$ is now required. The segment of g(t) viz., f(t), $t\in[0,1)$, and $\overleftarrow{f}(t)$, $t\in[-\tau,0)$ may separately be expanded in the normal intervals $[0,1)$ and $[-1,0)$ respectively in Walsh series as shown in Figures 2.9 (b) and 2.9(d). If

$$f(t-\tau) = \underline{f}_W^T \, \underline{D}_W^N \, \underline{w}(t), \quad t\in[0,1),$$

and $\overleftarrow{f}(t) = \overleftarrow{\underline{f}}_W^T \, \hat{\underline{w}}(t)$, $t\in[-1,0)$, where $\hat{\underline{w}}(t)$ is defined on $[-1,0)$, then the spectrum of $\overleftarrow{f}(t)$ for carry over into the interval $[0,1)$ is given by

$$\overleftarrow{\underline{f}}_{W_C} = \overleftarrow{\underline{f}}_W^T (\underline{D}_W^{m-N})^T \, .$$

Thus, the Walsh spectrum of $g(t-\tau)$ over $[0,1)$ including these effects of initial function is given by

$$\underline{g}_W^T = \underline{f}_W^T \, \underline{D}_W^N + \overleftarrow{\underline{f}}_W^T (\underline{D}_W^{m-N})^T \, . \tag{2.34}$$

2.4.7. Walsh operational matrix for stretch [W43(b)]

If the set of WF in $\underline{w}(t)$ on $[0,1)$ is stretched on the time scale such that the argument t changes to t/λ , where $\lambda>1$, we have $\underline{w}(t/\lambda) = \underline{S}_W \, \underline{w}(t)$, in which $\underline{S}_W$ is the Walsh stretched matrix (WSM). Let $\lambda=1+\kappa$. Stretch leads to expulsion of some part of the signal near t=1 from the considered interval. It is convenient to take a situation in which the signal over $\left[\frac{m-1}{m},1\right]$, the last segment, is completely expelled out of the interval $[0,1)$. We refer to this situation as that of single segment expulsion (SSE), under the condition that $\kappa=\frac{1}{m-1}$. If $\kappa<\frac{1}{m-1}$, the expulsion of the signal over the

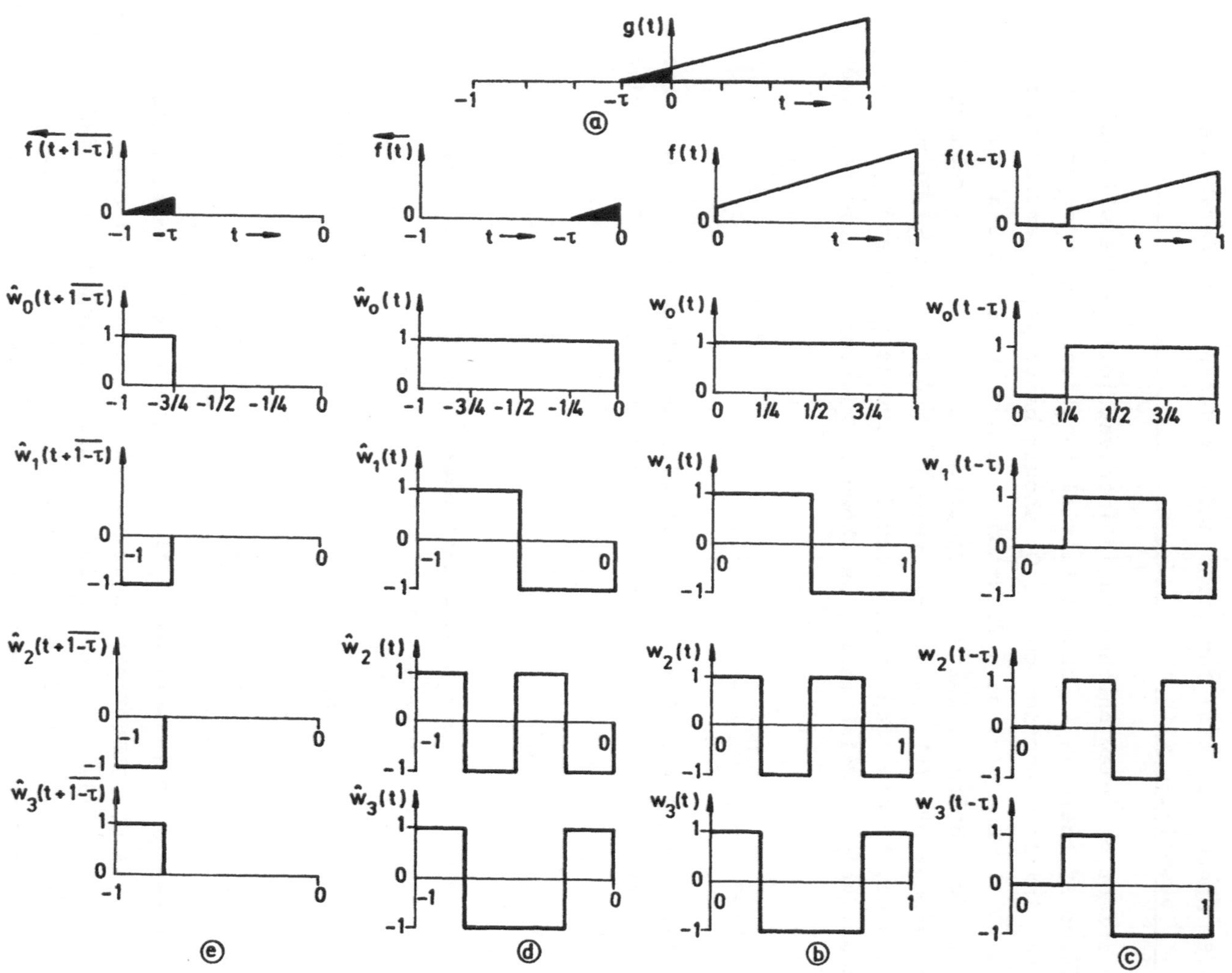

<u>Fig. 2.9.</u> Carry-over considerations in the Walsh spectrum of a delayed signal

last segment is partial.

By utilising the one-to-one correspondence of WF with BPF, we can generate the WSM for a given λ. By choosing $\Omega = \lambda/(\lambda-1)$, for SSE, the BPF stretch matrix $\underline{S}_\beta$ attains the form:

$$\underline{S}_\beta = \begin{bmatrix} \Omega-1 & 1 & & & 0 \\ & \Omega-2 & 2 & & \\ & & & & \\ & 0 & & 1 & \Omega-1 \\ & & & 0 & 0 \end{bmatrix} . \tag{2.35}$$

For an increased dimension, say $m=J.\,\Omega$ for any integer J, we get S_β the BPF stretch matrix using S'_β as

$$\underline{S}_\beta = \left[\begin{array}{cccc} \underline{S}'_\beta & & & \\ & \underline{S}'_\beta & & \\ & & \ddots & \\ & & & \underline{S}'_\beta \\ \hline & & 0 & \end{array}\right] \begin{array}{l} \left.\vphantom{\begin{array}{c}1\\1\\1\\1\end{array}}\right\} (\Omega-1)J \text{ rows} \\ \} \text{ J-rows} \end{array} . \tag{2.36}$$

$\underline{S}'_\beta$ is $\underline{S}_\beta$ with its last all-zero row dropped.

We then have

$$\underline{S}_W = \frac{1}{m} \underline{T}_{BW} \, \underline{S}_\beta \, \underline{T}_{BW}^{-1} . \tag{2.37}$$

Equation (2.36) has an important role to play in our future applications in simplifying computations. If $\lambda=1.25$

$$\underline{S}_{W(4x4)} = \frac{1}{4} \begin{bmatrix} 4 & 0 & 0 & 0 \\ 1 & 3 & 1 & -1 \\ 1 & 0 & 1 & 2 \\ -1 & 2 & 1 & 2 \end{bmatrix} . \tag{2.38}$$

In the case of SSE we have

$$\underline{S}^1_{W(2x2)} = \frac{1}{2} \begin{bmatrix} 2 & 0 \\ 2 & 0 \end{bmatrix} , \tag{2.39}$$

$$\underline{S}_{W(4x4)}^{1} = \frac{1}{3}\left[\begin{array}{c|c} \underline{S}_{W(2x2)}^{1}+2\underline{I}_{(2x2)} & \underline{S}_{W(2x2)}^{1}-2\underline{I}_{(2x2)} \\ \hline 2\underline{I}_{(2x2)}-\underline{S}_{W(2x2)}^{1} & \underline{M}_{(2x2)}-\underline{S}_{W(2x2)}^{1} \end{array}\right], \qquad (2.40)$$

$$\underline{S}_{W(8x8)}^{1} = \frac{1}{8}\left[\begin{array}{c|c} 3\underline{S}_{W(4x4)}^{1}+4\underline{I}_{(4x4)} & 3\underline{S}_{W(4x4)}^{1}-3\underline{I}_{(4x4)} \\ \hline \underline{I}_{(4x4)}-\underline{S}_{W(4x4)}^{1} & \underline{M}_{(4x4)}-\underline{S}_{W(4x4)}^{1} \end{array}\right], \qquad (2.41)$$

and so on. In the above matrices

$$\underline{M}_{(2x2)} = \begin{bmatrix} 1 & 2 \\ 2 & 1 \end{bmatrix}, \qquad (2.42a)$$

$$\underline{M}_{(4x4)} = \left[\begin{array}{cc|cc} 3 & 4 & 2 & 0 \\ 4 & 3 & 0 & 2 \\ \hline 2 & 0 & 3 & 4 \\ 0 & 2 & 4 & 3 \end{array}\right], \qquad (2.42b)$$

etc.

The superscript denotes the numbers of segments expelled. For $\lambda=1.25$,

$$\underline{S}_{W(8x8)}^{1} = \frac{1}{8}\left[\begin{array}{cccccccc} 8 & 0 & 0 & 0 & 0 & 0 & 0 & 0 \\ 2 & 6 & 2 & -2 & 2 & -2 & 2 & -2 \\ 2 & 0 & 2 & 4 & 0 & 2 & -4 & 2 \\ -2 & 4 & 2 & 4 & 0 & 2 & 0 & -2 \\ \cdots & \cdots & \cdots & \cdots & \cdots & \cdots & \cdots & \cdots \\ 0.5 & 1.5 & 0.5 & -0.5 & -1.5 & 3.5 & 0.5 & 3.5 \\ -0.5 & 2.5 & -2.5 & 2.5 & 1.5 & 0.5 & 1.5 & 2.5 \\ -0.5 & -1.5 & 5.5 & -1.5 & 1.5 & 0.5 & 1.5 & 2.5 \\ 0.5 & -2.5 & 0.5 & 3.5 & 2.5 & -0.5 & 4.5 & -0.5 \end{array}\right]. \qquad (2.43)$$

Notice that $\underline{S}_{W}^{1}$ of dimension 2^{k} is contained in one of dimension 2^{k+1}.

Direct generation of WSM

Fig. 2.10 shows the first four WF stretched with $\lambda=1+\kappa$. Expanding each of the stretched WF's in terms of the original set and rearranging

$$\underline{S}_{W(4x4)} = \frac{1}{4}\begin{bmatrix} 4 & 0 & 0 & 0 \\ 4\kappa & 4-4\kappa & 4\kappa & -4\kappa \\ 4\kappa & 0 & 4-12\kappa & 8\kappa \\ -4\kappa & 8\kappa & 4\kappa & 4-8\kappa \end{bmatrix} . \tag{2.44}$$

2.5. Operations and operational matrices via Haar functions

All the seven aspects discussed in Sections 2.4.1 to 2.4.7 may be developed with reference to HF. In the existing literature this does not seem to have been published so far. This is probably because such an exercise is merely of academic interest. We, therefore, will not take up this task here; but point out to the important transformations useful in this context, namely,

$$\underline{E}_h = \underline{T}_{BH}\, \underline{E}_\beta\, \underline{T}_{BH}^{-1} ,$$

$$\underline{D}_h = \underline{T}_{BH}\, \underline{D}_\beta\, \underline{T}_{BH}^{-1}, \text{ and} \tag{2.45}$$

$$\underline{S}_h = \underline{T}_{BH}\, \underline{S}_\beta\, \underline{T}_{BH}^{-1} .$$

Further details of the case of Haar functions are left to the reader as an exercise, which may well begin with integration of HF as indicated in Fig. 2.11 for the first four of the system of HF. The operational matrix E_h may be seen to be

$$\underline{E}_{h(2x2)} = \begin{bmatrix} \frac{1}{2} & -\frac{1}{4} \\ \frac{1}{4} & 0 \end{bmatrix} .$$

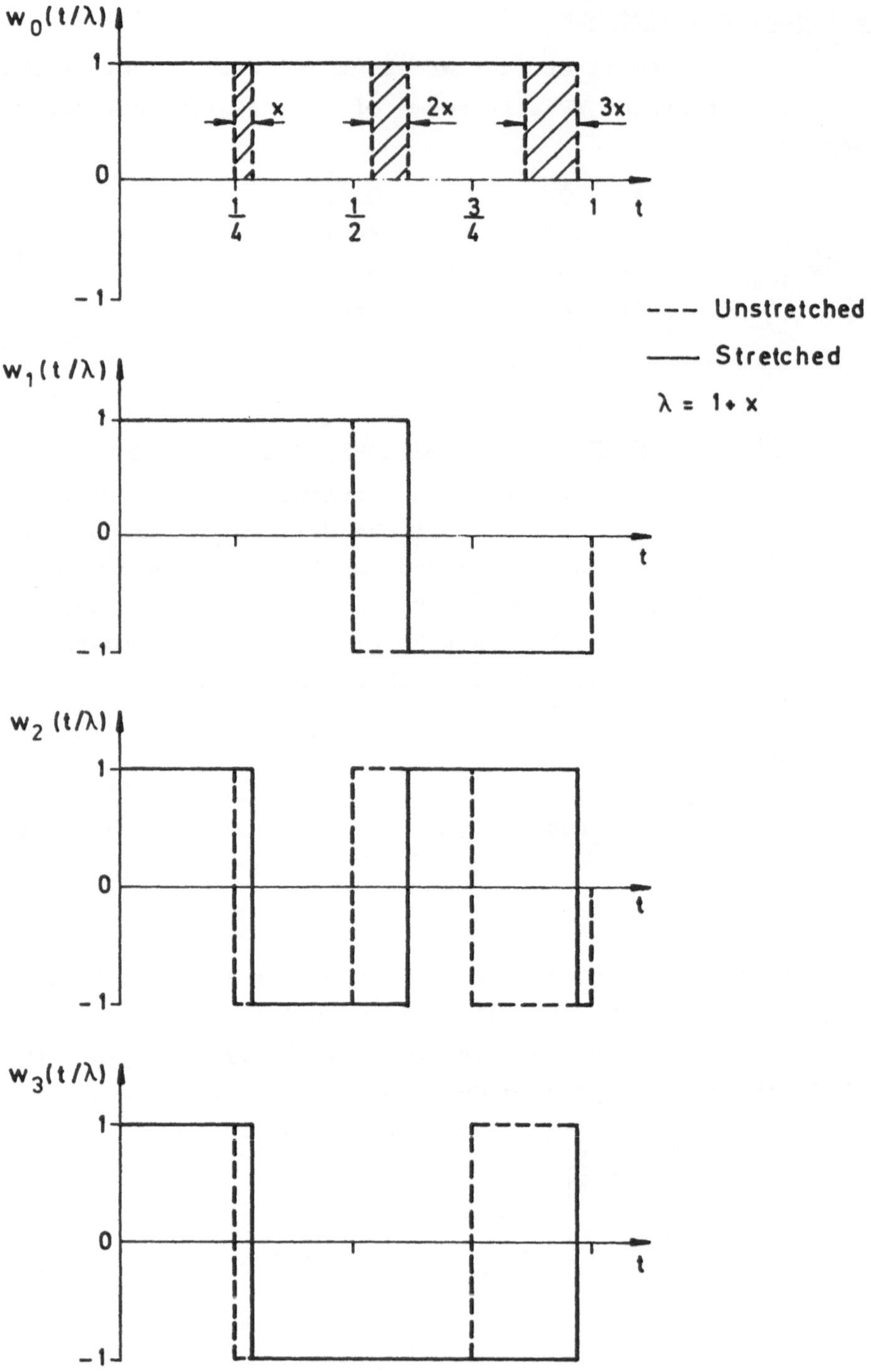

Fig. 2.10. Walsh functions stretched by $\lambda=1+x$

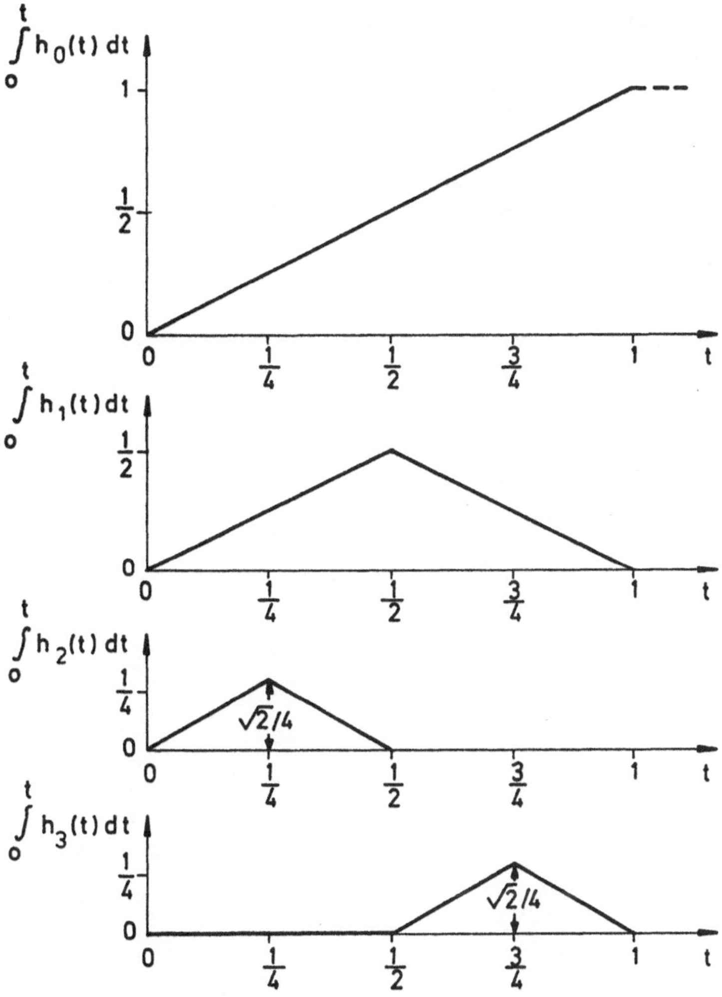

Fig. 2.11. Integrals of a set of four Haar functions

In general,

$$\underline{E}_{h\,(m\times m)} = \left[\begin{array}{c|c} \underline{E}_{h\,(\frac{m}{2}\times\frac{m}{2})} & -c\ \underline{T}_{BH\,(\frac{m}{2}\times\frac{m}{2})} \\ \hline c\ \underline{T}^{T}_{BH\,(\frac{m}{2}\times\frac{m}{2})} & \underline{O}_{(\frac{m}{2}\times\frac{m}{2})} \end{array}\right] , \qquad (2.46)$$

where

$$m = 2^k,$$

$$c = 2^{-\frac{3k+1}{2}}.$$

Analysis of Lumped Continuous Linear Systems

Continuous-time linear lumped dynamical systems are modelled in several different forms. The most widely used model is the description in state space in the form:

$$\frac{d\underline{x}(t)}{dt} = \underline{A}\ \underline{x}(t) + \underline{B}\ \underline{u}(t)$$

$$\underline{y}(t) = \underline{C}\ \underline{x}(t), \quad \underline{x}(0) = \underline{x}_o$$

The matrices $\underline{A}$, $\underline{B}$ and $\underline{C}$ may be time-varying, but we will, at this stage, restrict our attention to the time invariant case. $\underline{x}(t)$ is a n-vector of state, $\underline{u}(t)$ a p-vector of inputs, $\underline{y}$ is a q-vector of outputs and the matrices $\underline{A}$, $\underline{B}$ and $\underline{C}$ are of appropriate dimensions. We will begin with the solution of the problem in the simple, yet important, form

$$\frac{d\underline{x}}{dt} = \underline{A}\ \underline{x} + \underline{B}\ \underline{u}, \quad \underline{x}(0) = \underline{x}_o. \tag{3.1}$$

Let us expand $\frac{d\underline{x}}{dt}$, $\underline{x}$, and $\underline{B}\ \underline{u}$ in terms of an m-set of general PCBF, $\{\theta_i(t)\}$, as

$$\left.\begin{aligned} \frac{d\underline{x}(t)}{dt} &= \underline{V}\ \underline{\theta}(t), \\ \underline{x}(t) &= \underline{X}\ \underline{\theta}(t), \text{ and} \\ \underline{B}\ \underline{u}(t) &= \underline{P}\ \underline{\theta}(t), \end{aligned}\right\} \tag{3.2}$$

where $\underline{V}$, $\underline{X}$ and $\underline{P}$ are nxm matrices defined as

$$\left.\begin{aligned} \underline{V} &= [\,\underline{v}_1 \mid \underline{v}_2 \mid \cdots\ \cdots \mid \underline{v}_m], \\ \underline{v}_i &= [\,v_{1i} \mid v_{2i} \mid \cdots \mid v_{ni}]^T, \\ \underline{X} &= [\,\underline{x}_1 \mid \underline{x}_2 \mid \cdots\ \cdots \mid \underline{x}_m], \\ \underline{x}_i &= [\,x_{1i} \mid x_{2i} \mid \cdots \mid x_{ni}]^T, \\ \underline{P} &= [\,\underline{p}_1 \mid \underline{p}_2 \mid \cdots\ \cdots \mid \underline{p}_m], \\ \underline{p}_i &= [\,p_{1i} \mid p_{2i} \mid \cdots\ \cdots \mid p_{ni}]^T, \text{ and} \\ \underline{\theta}(t) &= [\,\theta_1(t) \mid \theta_2(t) \mid \cdots\ \cdots \mid \theta_m(t)]^T. \end{aligned}\right\} \tag{3.3}$$

Inserting (3.2) in (3.1),

$$\underline{V}\ \underline{\Theta}(t) = \underline{A}\ \underline{X}\ \underline{\Theta}(t) + \underline{P}\ \underline{\Theta}(t)\ . \tag{3.4}$$

Expressing $\underline{X}$ in terms of $\underline{V}$,

$$\underline{V} = \underline{A}\underline{V}\underline{E} + \hat{\underline{P}}, \tag{3.5}$$

where $\hat{\underline{P}} = [\hat{\underline{p}}_1\ \hat{\underline{p}}_2\ \cdots\ \hat{\underline{p}}_m]$,

$\hat{\underline{P}}$ = PCBF spectrum of $\{\underline{A}\ \underline{x}_o\} + \underline{P}$, and $\underline{A}\ \underline{x}_o$ is treated as a vector step function.

The algebraic equation (3.5) may be solved for the rate variable matrix $\underline{V}$ as a single long column $\mathcal{V}$ using the Kronecker product formula

$$\mathcal{V}_{(mnx1)} = [\underline{I}_{(mnxmn)} - \underline{E}^T_{(mxm)} \otimes \underline{A}_{(nxn)}]^{-1}\, \mathcal{P}_{(mnx1)}\ , \tag{3.6}$$

where

$$\mathcal{V}_{(mnx1)} = \begin{bmatrix} \underline{v}_1 \\ \underline{v}_2 \\ \vdots \\ \underline{v}_m \end{bmatrix}\ , \text{ and} \tag{3.7}$$

$$\mathcal{P}_{(mnx1)} = \begin{bmatrix} \hat{\underline{p}}_1 \\ \hat{\underline{p}}_2 \\ \vdots \\ \hat{\underline{p}}_m \end{bmatrix}\ . \tag{3.8}$$

From the above, we get the PCBF solution of (3.1) as

$$\underline{X}\ \underline{\Theta}(t) = \underline{V}\ \underline{E}\ \underline{\Theta}(t) + \{\text{PCBF spectrum of } \underline{x}_o\}\ \underline{\Theta}(t), \tag{3.9}$$

where $\underline{x}_o$ is taken as a vector step function.

When the basis is $\underline{\beta}(t)$, we take all BPF expansions distinguished by a subscript β. That is, we use $\underline{V}_\beta$, $\underline{X}_\beta$, $\underline{P}_\beta$, $\mathcal{V}_\beta$, $\mathcal{P}_\beta$ defined by

$$V_\beta = [\underline{v}_{\beta_1} \mid \underline{v}_{\beta_2} \mid \cdots \mid \underline{v}_{\beta_m}] , \tag{3.10a}$$

$$\underline{v}_{\beta_i} = [v_{\beta_{1i}}\ \ v_{\beta_{2i}}\ \cdots\ v_{\beta_{ni}}]^T, \tag{3.10b}$$

$$X_\beta = [\underline{x}_{\beta_1} \mid \underline{x}_{\beta_2} \mid \cdots \mid \underline{x}_{\beta_m}] , \tag{3.10c}$$

$$\underline{x}_{\beta_i} = [x_{\beta_{1i}}\ \ x_{\beta_{2i}}\ \cdots\ x_{\beta_{ni}}]^T, \tag{3.10d}$$

$$\underline{P}_\beta = [\underline{p}_{\beta_1} \mid \underline{p}_{\beta_2} \mid \cdots \mid \underline{p}_{\beta_m}] , \tag{3.10e}$$

$$\underline{p}_{\beta_i} = [p_{\beta_{1i}}\ \ p_{\beta_{2i}}\ \cdots\ p_{\beta_{ni}}]^T, \tag{3.10f}$$

$$\hat{\underline{P}}_\beta = [\underline{A}\ \underline{x}_o\ \ \underline{A}\ \underline{x}_o\ \cdots\ \underline{A}\ \underline{x}_o] + \underline{P}_\beta , \tag{3.10g}$$

$$\hat{\underline{p}}_{\beta_i} = \underline{p}_{\beta_i} + \underline{A}\ \underline{x}_o , \tag{3.10h}$$

$$\mathcal{V}_\beta = \begin{bmatrix} \underline{v}_{\beta_1} \\ \hline \underline{v}_{\beta_2} \\ \hline \vdots \\ \hline \underline{v}_{\beta_m} \end{bmatrix} , \tag{3.10i}$$

$$\mathcal{P}_\beta = \begin{bmatrix} \hat{\underline{p}}_{\beta_1} \\ \hline \hat{\underline{p}}_{\beta_2} \\ \hline \vdots \\ \hline \hat{\underline{p}}_{\beta_m} \end{bmatrix} . \tag{3.10j}$$

The operational matrix $\underline{E}_\beta$ will replace $\underline{E}$. Thus, the BPF solution of (3.1) is given by

$$\underline{X}_\beta \underline{\beta}(t) = \underline{V}_\beta \underline{E}_\beta\ \underline{\beta}(t) + \underbrace{[\underline{x}_o\ \ \underline{x}_o\ \cdots\ \underline{x}_o]}_{m\ \text{col.}}\underline{\beta}(t)\,. \tag{3.11}$$

While using WF as the basis, we distinguish respective spectra with the subscript 'w'. That is,

$$V_w = [\underline{v}_{w_o} \mid \underline{v}_{w_1} \mid \cdots \mid \underline{v}_{w_{m-1}}]\ , \tag{3.12a}$$

$$\underline{v}_{w_i} = [v_{1i}\ \ v_{w_{2i}}\ \ \ v_{w_{ni}}]^T\ , \tag{3.12b}$$

$$X_w = [\underline{x}_{w_o} \mid \underline{x}_{w_1} \mid \cdots \mid \underline{x}_{w_{m-1}}]\ , \tag{3.12c}$$

$$\underline{x}_{w_i} = [x_{w_{1i}}\ \ x_{w_{2i}}\ \ \ x_{w_{ni}}]^T\ , \tag{3.12d}$$

$$\underline{P}_w = [\underline{p}_{w_o} \mid \underline{p}_{w_1} \mid \cdots \mid \underline{p}_{w_{m-1}}]\,, \tag{3.12e}$$

$$\underline{p}_{w_i} = [p_{w_{1i}}\ \ p_{w_{2i}}\ \ \ p_{w_{ni}}]^T\ , \tag{3.12f}$$

$$\hat{\underline{P}}_w = [\underline{A}\ x_o\ \ \underbrace{0\ \ 0\ \ \cdots\ \ 0}_{m-1\ \text{col.}}] + \underline{P}_w\ , \tag{3.12g}$$

$$= [\hat{\underline{p}}_{w_o} \mid \hat{\underline{p}}_{w_1} \mid \cdots \mid \hat{\underline{p}}_{w_{m-1}}]\text{, say,} \tag{3.12h}$$

$$\mathcal{V}_w = \begin{bmatrix} \underline{v}_{w_o} \\ \hdashline \underline{v}_{w_1} \\ \hdashline \vdots \\ \hdashline \underline{v}_{w_{m-1}} \end{bmatrix}, \tag{3.12i}$$

$$\mathcal{P}_w = \begin{bmatrix} \hat{\underline{p}}_{w_o} \\ \hline \hat{\underline{p}}_{w_1} \\ \hline \vdots \\ \hline \hat{\underline{p}}_{w_{m-1}} \end{bmatrix} . \tag{3.12j}$$

Inserting $\underline{E}_w$ in the appropriate place, the Walsh spectral solution of (3.1) is

$$\underline{X}_w \, \underline{w}(t) = \underline{V}_w \underline{E}_w \underline{w}(t) + [\underline{x}_o \; \underbrace{0 \; 0 \; \ldots \; 0}_{(m-1)\text{col.}}] \, \underline{w}(t) . \tag{3.13}$$

The Kronecker product formula (3.6) is computationally very laborious. Chen and Hsiao's [W11] attempt to reduce the computational effort, avoiding direct inversion of (mnxmn)size matrices is also considerably tedious and complicated. This Kronecker product reduction method still requires at least k ($=\log_2 m$) inversions of (nxn) matrices, 3k multiplications of (nxn) matrix pairs, and $\frac{k(k-1)}{2} + 2^k$ multiplications of an (nxn) matrix with an n-vector [W57]. In practice, there may arise the following situations necessitating inclusion of large number of terms in each expansion with corresponding rise in the computational burden in the algorithm.

i) To increase the accuracy of solution in the initial normal interval [0,1) itself.

ii) To compute the solution of (3.1) for large t, beyond the naturally normal initial interval [0,1) maintaining the same accuracy as in this initial interval.

In the case of large intervals of time, computations have to be made with increased m to maintain accuracy, if the whole period is normalised by time scaling, bringing the problem to the standard form.

iii) To compute the solution of (3.1) in an extended interval, say [1,2) for instance, while the solution in [0,1) is available a priori in the form of (3.13).

Normalisation of [0,2) simultaneously doubling the value of m to maintain the same accuracy, would lead us to the present standard algorithm. But we do not know how to use the solution available already to reduce the computational burden.

iv) To gain insight into the nature of propagation of errors and stability of the resulting computations.

Chen and Hsiao's [W11] reduction recommends an inevitable increase in m for i) and ii). It neither suggests a means of utilising the available solution in case iii) nor does it provide any insight into the stability of errors in case iv). We now discuss a technique of continuing solution [W45] of (3.1) across contiguous normal (or normalised) intervals of time, ultimately attending to the above four practically important situations in the Walsh series analysis of (3.1).

3.1. Continuing solution beyond the limit of the initial normal interval [W45]

It is clear by now that the BPF coefficients represent the average values of the function. We now need the average values. From the Walsh spectrum these can be obtained through the linear orthogonal transformation

$$\underline{V}_\beta = [\underline{v}_{\beta 1} \ \underline{v}_{\beta 2} \ \cdots \ \underline{v}_{\beta m}] = \underline{T}_{BW}\underline{V}_W \quad . \tag{3.14}$$

At the end of the interval [0,1), i.e., at t=1, the state is given by

$$\underline{x}(1) = \frac{1}{m} \sum_{i=1}^{m} \underline{v}_{\beta_i} + \underline{x}(0) \quad . \tag{3.15}$$

Having obtained $\underline{x}(1)$ from (3.15), we are now in a position to compute the solution in the next normal interval [1,2), by defining WF appropriately in this interval. The situation is as in (3.1) but with a change of the origin from t=0 to t=1.

We are thus in a position to write a useful recurrence formula connecting Walsh series solutions in two contiguous unit intervals [j-1,j) and [j,j+1)

$$\left.\begin{aligned}
\hat{\underline{P}}_W^{(j)} &= [\underline{A}\ \underline{x}(j-1)\ \ \underbrace{0\ 0\ \ldots\ 0}_{(m-1)\,\text{col.}}] + \underline{P}_W^{(j)} \ , \\
\mathcal{V}_W^{(j)} &= [\underline{I} - \underline{E}_W^T \otimes \underline{A}]^{-1} \mathcal{P}_W^{(j)} \ , \\
\underline{x}(j) &= \frac{1}{m} \sum_{i=1}^{m} \underline{v}_{\beta_i}^{(j)} + \underline{x}(j-1),\ j=1,2,3,\ldots, \\
\underline{x}(j) &= \underline{x}(t)\Big|_{t=j} \ .
\end{aligned}\right\} \qquad (3.16)$$

The superscript j on the terms in (3.16) means that they belong to the j-th unit interval.

3.2. Continuation with normalised single segment approximation (m=1)

To maximize the reduction of computational effort in the algorithm, we would naturally choose the single segment approximation. To retain the specified level of accuracy, we maintain the absolute time width of the segment unaltered. That is, an interval $[0, \frac{1}{m})$ is scaled to unit length in which we need to consider only the first term of the Walsh series. We will therefore call this technique as the single term Walsh series (STWS) method. As a result of the changed time scale, the system matrices in (3.1) will now become $\frac{\underline{A}}{m}$ and $\frac{\underline{B}}{m}$ respectively. Furthermore, $\underline{E}_W = \frac{1}{2}$, $\mathcal{V}_W = \underline{V}_W = \underline{v}_{W_0}$, $X_W = \underline{x}_{W_0}$ and $\mathcal{P}_W = \underline{P}_W = \hat{\underline{P}}_{W_0}$.

$\underline{V}_W$, $\underline{X}_W$ and $\hat{\underline{P}}_W$ thus reduce to n-vectors. Equation (3.16) now takes the form:

$$\left.\begin{aligned}
\underline{V}_W^{(j)} &= (\underline{I} - \frac{\underline{A}}{2m})^{-1}\ [\underline{P}^{(j)} + \frac{\underline{A}}{m}\underline{x}(j-1)] \ , \\
\underline{X}_W^{(j)} &= \underline{V}_W^{(j)} \underline{E}_W + \underline{x}(j-1), \\
\underline{x}(j) &= \underline{x}(j-1) + \underline{V}_W^{(j)} \ , \quad j=1,2,\ldots \ .
\end{aligned}\right\} \qquad (3.17)$$

Solution by (3.17) on the whole involves only one inversion of a nxn matrix, one multiplication of a pair of nxn matrices and 2m multiplications of an nxn matrix with a n-vector. The third in equations (3.17) gives us the state at the junctions of the segments as a bonus. In addition, we have freedom to compute over any number of segments m, without the restriction of $m=2^k$.

Example 3.1.

Consider equation (3.1) with

$$\underline{A} = \begin{bmatrix} -1 & -1.8 \\ 5 & -1 \end{bmatrix}, \quad \underline{B} = \begin{bmatrix} 1.8 \\ 0 \end{bmatrix}, \quad \underline{x}(0) = \begin{bmatrix} 0 \\ 0 \end{bmatrix}$$

and u(t) = unit step. With m=4, Table 3.1 shows the results by the two methods discussed above.

Time in sec.	Solution by the present method		By the Kronecker product formula
	$\underline{x}^T$	$\underline{x}^T$	$\underline{x}^T$
0		(0.0 0.0)	
	(0.18 0.099999)		(0.1800415 0.1000001)
$\frac{1}{4}$		(0.360 0.199998)	
	(0.436000 0.419975)		(0.4359585 0.4199999)
$\frac{1}{2}$		(0.512 0.639998)	
	(0.4872036 0.8395557)		(0.4871585 0.8395553)
$\frac{3}{4}$		(0.4624072 1.0391194)	
	(0.383667 1.1368101)		(0.3837036 1.1367999)
1.0		(0.304926 1.2345008)	

Table 3.1. Solution of Example 3.1.

The numerous matrix inversions in the Kronecker product formula give rise to round off errors that can be reduced only by working with double precision arithmetic. This is not the case with the STWS method which is simple, recursive and has the following advantages. Computations may be carried on to any length of time (once stability is ensured).
The solution known a priori on an initial section of the interval is efficiently utilised. The state at the transitions, $\underline{x}(j)$ can be computed. The most important advantage is the insight gained into the stability of computations. The sequence generated by (3.17) will be stable if the eigenvalues of $(\underline{I}-\underline{A}/2m)^{-1}(\underline{I}+\underline{A}/2m)$ lie inside the unit circle. If (3.1) is physically stable, the computations will always be stable if $(\underline{I}-\underline{A}/2m)^{-1}$ is ensured to exist.

The STWS method may be generally termed as the single term PCBF method, for, the technique of extension of computations is independent of the type of PCBF. After all, the first term of all the PCBF systems on a normal interval happens to be an unit pulse on [0,1), when m=1. The names $\beta_1(t)$, $w_o(t)$, and $h_o(t)$ become superficial. Consequently, equations (3.17) are the same as those for the BPF basis derived by Sannuti [W57]. The only difference is that the limit on the running index of the recursive sequence can be indefinitely lifted. The BPF solution (3.1) may be shown to be [W57]

$$\underline{X}_{\beta_i} = (\underline{I} - \frac{\underline{A}}{2m})^{-1} [\underline{x}_o + \frac{1}{2m} \underline{P}_{\beta_1}] ,$$

$$\underline{X}_{\beta_{i+1}} = (\underline{I} - \frac{\underline{A}}{2m})^{-1} (\underline{I} + \frac{\underline{A}}{2m}) \underline{X}_{\beta_i} + (\underline{I} - \frac{\underline{A}}{2m})^{-1} [\underline{P}_{\beta_{i+1}} + \underline{P}_{\beta_i}] \quad (3.18)$$

for all $i \geq 1$.

Notice that the STWS solution given by (3.17) and the BPF solution given by (3.18) are identical. The STWS solution gives the state at the instants $t_i = \frac{i-1}{m}$, i=1,2,... . This can be obtained from (3.18) also in the same manner.

At this stage, the following conclusions can be made:

i) The single term (Walsh or any other series) approach is the simplest from the point of view of computational economy.

ii) The extension technique [W45] makes computation possible for any length of time if desired.

iii) The single term approach provides an insight into the stability of computations.

iv) The single term approach with facility for continuation now makes it clear that the name of the series of PCBF becomes trivial.

v) Equations (3.17 and 3.18) are strikingly similar to the trapeziodal formula (modified Euler formula) for integrating (3.1). In fact, if $\underline{B}\ \underline{u}(t) = 0$, or when $\underline{B}\ \underline{u}(t)$ is itself approximated by trapezoidal rule into a PCBF term, the PCBF methods are identical with the well known trapezoidal formula.

vi) in view of v) above, the well known modified Euler method or trapezoidal rule get a fresh and sound sanction with the PCBF as the new basis of approximating orthogonal series.

The simplicity of solutions via PCBF for (3.1) culminating in an identity, under certain conditions, with some of the well established numerical methods of integration of differential equations, is not a surprising phenomenon. Furthermore, these features of simplicity and similarity with the existing well known methods do not imply that all the exercises on the applications of PCBF such as WF and BPF have turned out to be trivial. Equation (3.1) is but one of the many forms of mathematical models used in practice, and is the simplest time invariant case. If the matrices $\underline{A}$, $\underline{B}$ and $\underline{C}$ become time-varying or if the differential equations are bilinear or in other nonlinear forms, the above arguments on similarity with the well known numerical methods cannot be easily extended to such cases.

In the following chapters, we will develop algorithms for integrating some kinds of differential equations containing features such as delays, nonlinear terms, terms with stretched arguments etc. to make it clear that the PCBF approach to the solution of differental euqations has its own identity and advantages. Before leaving this chapter we will illustrate the superiority of OSOMRI in the solution of time-invariant linear systems.

Example 3.2. Consider a second order equation

$$\frac{d^2x}{dt^2} = -\pi^2 x(t), \quad x(0) = 0, \ \dot{x}(0) = -\pi \tag{3.19}$$

The exact solution is $x(t) = -\sin\pi t$. We will attempt to get the solution of (3.19) directly without transforming it into the form of (3.1). With m-size BPF series, for instance, on [0,1) the solution may be directly obtained from the following:

a) using E_β^2.

$$\underline{x}_\beta^T \underline{\beta}(t) = -\pi^2 \underline{x}_\beta^T \underline{E}_\beta^2 \underline{\beta}(t) + \underbrace{[-\pi \ -\pi \ \ldots \ -\pi]}_{m \text{ col.}} \underline{E}_\beta \underline{\beta}(t), \tag{3.20a}$$

b) using E_{β_2}

$$\overset{*}{\underline{x}}{}_\beta^T \underline{\beta}(t) = -\pi^2 \overset{*}{\underline{x}}{}_\beta^T \underline{E}_{\beta_2} \underline{\beta}(t) + \underbrace{[-\pi \ -\pi \ \ldots \ -\pi]}_{m \text{ col.}} \underline{E}_\beta \underline{\beta}(t), \tag{3.20b}$$

where $\underline{x}_\beta$ and $\overset{*}{x}_\beta$ denote vectors of BPF components of x(t) in the above two cases.

We distinguish the result of (3.20b) due to the OSOMRI $\underline{E}_{\beta_2}$ with the asterisk mark. $\underline{x}_\beta$ and $\overset{*}{\underline{x}}_\beta$ are computed and shown in Table 3.2 for m=16. Notice that

$$x_{\beta_1} = \frac{-2m\pi}{(4m^2+\pi^2)} = -\ 0.0972375 ,$$

$$\overset{*}{x}_{\beta_1} = \frac{-3m\pi}{(6m^2+\pi^2)} = -0.097549 .$$

These results, after rounding to four decimal places, are shown in Table 3.2. $\overset{*}{\underline{x}}_\beta$ is closer to the true solution. The solution is shown only upto t=1/2, the point of symmetry in [0,1).

Equations (3.20a and b) imply that

$$\underline{x}_\beta^T [I + \frac{\pi^2}{12m^2}(I + \pi^2 E_{\beta_2})^{-1}] = \overset{*}{\underline{x}}{}_\beta^T .$$

k	Piecewise constant sub-interval values		
	x_{β_k} (through $\underline{E}_\beta^2$)	$x^*_{\beta_k}$ (through $\underline{E}_{\beta_2}$)	Exact
1	-0.0972	-0.0975	-0.0979
2	-0.2880	-0.2889	-0.2898
3	-0.4676	-0.4691	-0.4706
4	-0.6293	-0.6313	-0.6334
5	-0.7668	-0.7693	-0.7718
6	-0.8749	-0.8777	-0.8805
7	-0.9493	-0.9523	-0.9554
8	-0.9872	-0.09904	-0.9936

Table 3.2. Solution of equation (3.19)

This relationship is free from initial condition terms and therefore applies to the constant subinterval values as well. That is,

$$\frac{x^*_{\beta_k}}{x_{\beta_k}} = 1 + \nu \text{ , where } \nu = \frac{\pi^2}{12m^2}\left(1 + \frac{\pi^2}{6m^2}\right).$$

In the present case, $\nu = 3.2334036 \times 10^{-3}$. The results in Table 3.2 may be seen to confirm this feature. Thus we see that in the solution of higher order problems by direct method, the OSOMRI is more accurate.

IV

Analysis of Time Delay Systems

System of equations with a single delay.

Consider a time delay system described by

$$\left.\begin{aligned} \dot{\underline{x}}(t) &= \underline{A}\ \underline{x}(t) + \underline{L}\ \underline{x}(t-\tau) + \underline{B}\ \underline{u}(t), \\ \underline{x}(t) &= \underline{x}_b(t),\ t \in [-\tau, 0), \end{aligned}\right\} \tag{4.1}$$

where $\underline{x}$ is a n-vector of state, $\underline{u}$ is an input vector of l components, $\underline{A}$, $\underline{L}$ and $\underline{B}$ are nxn, nxl matrices respectively. We will first establish a procedure to solve (4.1) with the help of operational matrices $\underline{D}$ and $\underline{E}$. Initially we get a solution over [0,1) by expanding $\dot{x}$, x and $\underline{B}\ \underline{u}$ in series of PCBF of size m as detailed in (3.2) and (3.3). In addition, we let

$$\underline{x}(t-\tau) = \underline{X}\ \underline{\Theta}(t). \tag{4.2}$$

The initial function $\underline{x}_b(t)$ will be expanded in PCBF series $\hat{\underline{\Theta}}(t)$, $t \in [-1, 0)$ as

$$\underline{x}_b(t) = \underline{X}_b \hat{\Theta}(t), \tag{4.3}$$

where

$$\underline{X}_b = [\underline{x}_{b_1} \mid \underline{x}_{b_2} \mid \dots \mid \underline{x}_{b_m}] \ .$$

Accordingly, $\underline{x}_{b_i}$ is the ith column of $\underline{X}_b$.
Inserting these expansions in (4.1)

$$\underline{V}\ \underline{\theta}(t) = \underline{A}\ \underline{X}\ \underline{\Theta}(t) + \underline{L}\ \underline{X}\ \underline{\Theta}(t) + \underline{P}\ \underline{\Theta}(t) \ . \tag{4.4}$$

Also

$$\underline{X}\ \underline{\theta}(t) = [\underline{X}\ \underline{D}^N + \underline{X}_b (\underline{D}^{n-N})^T] \underline{\Theta}(t) \ . \tag{4.5}$$

Expressing $\underline{X}$ in terms of $\underline{V}$ and using the operational matrix $\underline{E}$,

$$\underline{V}\,\underline{\theta}(t) = \underline{A}\,\underline{V}\,\underline{E}\,\underline{\Theta}(t) + [\text{PCBF spectrum of } \underline{A}\,x_o \text{ as a column}]\underline{\Theta}(t) + \underline{L}\,\underline{V}\,\underline{E}\,\underline{D}^N\,\underline{\Theta}(t) + [\text{PCBF spectrum of } \underline{L}\,x_o \text{ as a column}]\underline{\Theta}(t) + \underline{L}\,\underline{X}_b(\underline{D}^{m-N})^T\underline{\Theta}(t) + \underline{P}\,\underline{\Theta}(t) \;, \tag{4.6}$$

leading to

$$\underline{V} = \underline{A}\,\underline{V}\,\underline{E} + \underline{L}\,\underline{V}\,\underline{E}\,\underline{D}^N + \overset{*}{\underline{P}} \;,$$

where $\overset{*}{\underline{P}} = [\overset{*}{\underline{p}}_1 \,\vdots\, \overset{*}{\underline{p}}_2 \,\vdots\, \cdots \,\vdots\, \overset{*}{\underline{p}}_m]$,

$$\overset{*}{\underline{P}} = \underline{P} + \text{PCBF spectrum of } \underline{A}\,\underline{x}_o + \text{PCBF spectrum of } \underline{L}\,\underline{x}_o + \underline{L}\,\underline{X}_b(\underline{D}^{m-N})^T. \tag{4.7}$$

With the long column approach as in (3.6),

$$\mathcal{V} = [\underline{I} - \underline{A} \otimes \underline{E}^T - \underline{L} \otimes (\underline{E}\,\underline{D}^N)^T]^{-1}\overset{*}{\mathcal{P}} \;, \tag{4.8}$$

where

$$\overset{*}{\mathcal{P}} = [\overset{*}{\underline{p}}{}_1^T \,\vdots\, \overset{*}{\underline{p}}{}_2^T \,\vdots\, \cdots \,\vdots\, \overset{*}{\underline{p}}{}_m^T]^T .$$

Having obtained the PCBF spectrum of the rate variable, we get that of the state as

$$\underline{X} = \underline{V}\,\underline{E} + [\text{PCBF spectrum of } \underline{x}_o \text{ as a column}]. \tag{4.9}$$

<u>Particular cases</u>

a) BPF spectra of $\underline{A}\,\underline{x}_o$, L $\underline{x}_o$ and $\underline{x}_o$ occuring in equations (4.6), (4.7) and (4.9) are respectively as follows:

$$\underbrace{[\underline{A}\,\underline{x}_o \;\; \underline{A}\,\underline{x}_o \;\cdots\; \underline{A}\,\underline{x}_o]}_{m \text{ col.}} ,$$

$$[\underline{L}\,\underline{x}_o,\ \underline{L}\,\underline{x}_o,\ \ldots\ \underline{L}\,\underline{x}_o]\ ,\ \text{and}$$
$$\underbrace{\qquad\qquad\qquad}_{\text{m col.}}$$

$$\underbrace{[x_o,\ x_o,\ \ldots\ x_o]}_{\text{m col.}}\ .$$

With these spectra along with the respective matrices and vectors distinguished by the subscript 'β', the above analysis may be done with respect to BPF.

b) Walsh spectra of the terms $\underline{A}\,\underline{x}_o$, $\underline{L}\,\underline{x}_o$ and $\underline{x}_o$ will similarly be used as follows:

$$[\underline{A}\,\underline{x}_o,\ \underbrace{0,\ 0,\ \ldots\ 0}_{\text{m-1 col.}}]\ ,$$

$$[\underline{L}\,\underline{x}_o,\ \underbrace{0,\ 0,\ \ldots\ 0}_{\text{m-1 col.}}]\ ,$$

$$[\underline{x}_o,\ \underbrace{0,\ 0,\ \ldots\ 0}_{\text{m-1 col.}}]\ .$$

4.1. Continuing solution beyond the limit of the initial normal interval

Equations (4.2) to (4.9) establish a procedure to compute the solution of (4.1) on the interval $[0,1)$ taking into account the initial function terms on $[-\tau,0)$. To continue computations into the next interval $[1,2)$, we consider PCBF series over $[1,2)$ and require $\underline{X}$, the PCBF spectrum of the state in $[0,1)$ from which the spectrum corresponding to $[1-\tau,1)$ will be carried over into $[1,2)$. We also need the state $\underline{x}(1)$ calculated from (3.15).

We can write a recurrence formula for continuation from a general unit interval to the next i.e., $[k-1,k)$ to $[k,k+1)$.

$$\overset{*}{\underline{P}}^{(k)} = \{\{\underline{A}\ \underline{x}(k-1)\}\} + \{\{\underline{L}\ \underline{x}(k-1)\}\}\,\underline{D}^N + \underline{L}\ \underline{X}^{(k-1)}(\underline{D}^{m-N})^T + \underline{P}^{(k)}, \quad (4.10)$$

where $\{\{\ \}\}$ means "a matrix of m columns representing the PCBF spectrum of"

$$\mathcal{V}^{(k)} = [\underline{I} - \underline{A} \otimes \underline{E}^T - \underline{L} \otimes (\underline{E}\ \underline{D}^N)^T]^{-1} \overset{*}{\mathcal{P}}^{(k)}, \quad (4.11)$$

$$x(k) = \frac{1}{m} \sum_{i=1}^{m} \underline{v}^{(k)}_{\beta_i} + \underline{x}(k-1), \quad (4.12)$$

$$\underline{x}(k) = \underline{x}(t)\Big|_{t=k},$$

$k = 1,2,3,\ldots$.

The superscript $^{(k)}$ signifies that the quantity belongs to the k-th normal interval.

4.2. Continuation with normalised single segment approximation (m=1)

In the case $\tau = \frac{N}{m}$, $\underline{D}^1_{(1)} = 0\ \underline{D}^0_{(1)} = \underline{I}_{(1x1)} = 1$. Also $\underline{E} = 1/2$ and the carry-over spectrum is $\underline{X}^{[k-N)}$. With a changed scale we stretch the segment of length $\frac{1}{m}$ to be normal. Equation (4.8) reduces to the single column form

$$\underline{V}^{(k)} = [1 - \underline{A}/2m]^{-1} [\frac{\underline{A}}{m}\underline{x}^{(k-1)} + \frac{\underline{L}}{m} \underline{X}^{[k-N)} + \underline{P}^{(k)}] . \quad (4.13)$$

Also

$$\underline{X}^{(k)} = \frac{1}{2} \underline{V}^{(k)} + \underline{x}(k-1), \text{ and} \quad (4.14)$$

$$\underline{x}(k) = \underline{V}^{(k)} + \underline{x}(k-1) . \quad (4.15)$$

We have deliberately dropped the subscripts β, w etc. from the above because the first components in these series $\underline{\beta}(t)$ or $\underline{w}(t)$ are not different.

4.3. Single term PCBF approach for systems with multiple delays

Consider a system described by

$$\underline{\dot{x}} = \underline{A}\,\underline{x} + \sum_{s=1}^{Q} \underline{L}_s\,\underline{x}(t-\tau_s) + \underline{B}\,\underline{u}(t-\tau_k), \tag{4.16}$$

where

$$\tau_1 < \tau_2 < \tau_3 \; ...< \tau_Q \quad , \tag{4.17}$$

$$\left.\begin{array}{l} \underline{x}(t) = \underline{x}_b(t), \; -\tau_s \leq t < 0 \\ \underline{u}(t) = \underline{u}_b(t), \; -\tau_u \leq t < 0 \end{array}\right\} . \tag{4.18}$$

Changing the time scale such that $\hat{t} = m\cdot t$,

$$\underline{\dot{x}}(\hat{t}) = \frac{\underline{A}}{m}\,\underline{x}(\hat{t}) + \sum_{s=1}^{Q} \frac{\underline{L}_s}{m}\,\underline{x}[\hat{t} - (N_s+\alpha_s)] + \frac{1}{m}\,\underline{B}\,\underline{u}[\hat{t} - (N_u+\alpha_u)]. \tag{4.19}$$

In the above, $\tau_s = (N_s+\alpha_s)/m$ and $\tau_u = (N_u+\alpha_u)/m$. N represents the integer part of the delay and α the fractional part. Equation (4.19) may be solved by the following single term PCBF formula:

$$\underline{V}^{(k)} = (\underline{I}-\frac{\underline{A}}{2m})^{-1}\{\frac{\underline{A}}{m}\underline{x}(k-1) + \sum_{s=1}^{Q} \frac{\underline{L}_s}{m}\,[(1-\alpha_s)\,\underline{X}^{(k-N_s)} + \alpha_s\underline{X}^{(k-N_s-1)}]$$

$$+ \frac{1}{m}\,[(1-\alpha_u)\underline{P}^{(k-N_u)} + \alpha_u\underline{P}^{(k-N_u-1)}]\} , \tag{4.20}$$

$$\underline{X}^{(k)} = \frac{1}{2}\,\underline{V}^{(k)} + \underline{x}(k-1) , \tag{4.21a}$$

$$\underline{x}(k) = \underline{V}^{(k)} + \underline{x}(k-1) , \quad k=1,2,3,... . \tag{4.21b}$$

With WF as the basis, analysis of delay systems has been carried out in [W36]. Prasada Rao and Srinivasan [W52, W53] employ BPF in the situation.

4.4. Illustrative examples

Example 1: Consider a unity feedback system [D7] as shown in Fig.4.1.

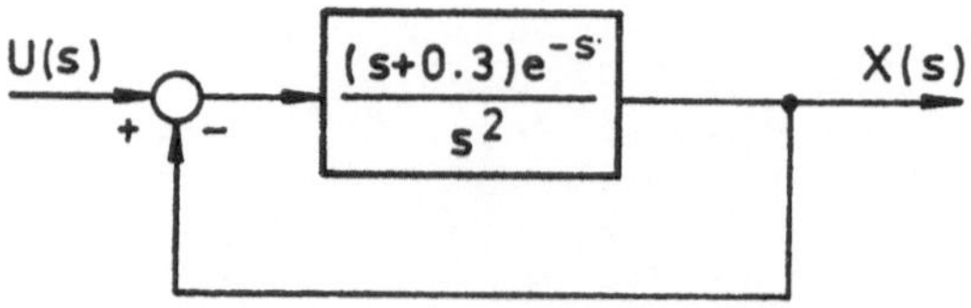

Fig. 4.1. A unity feedback system with delay [D7]

The closed loop transfer function of the system

$$\frac{X(s)}{U(s)} = \frac{(s+0.3)\exp(-s)}{s^2+(s+0.3)\exp(-s)} \, . \tag{4.22}$$

The differential equation is

$$\ddot{x}(t) + \dot{x}(t-1) + 0.3x(t-1) = \dot{u}(t-1) + 0.3k(t-1) \, , \tag{4.23}$$

$$\left.\begin{aligned} \dot{x}(t) &= 0 \\ x(t) &= 0 \end{aligned}\right\} \, , \quad t\in[-1,0] \, ,$$

where u(t) is a ramp function delayed by 1 sec. Equation (4.23) may be written as

$$\dot{\underline{x}}(t) = \begin{bmatrix} 0 & 1 \\ 0 & 0 \end{bmatrix} \underline{x}(t) + \begin{bmatrix} 0 & 0 \\ -0.3 & -1 \end{bmatrix} \underline{x}(t-1) + \begin{bmatrix} 0 \\ \dot{u}(t-1)+0.3u(t-1) \end{bmatrix} \, , \tag{4.24}$$

and $\underline{x}(t) = \begin{bmatrix} 0 \\ 0 \end{bmatrix}$, $t\in[-1,0]$.

We take m=1, N=1 and get

$$(\underline{I} - \frac{A}{2})^{-1} = \begin{bmatrix} 1 & 1/2 \\ 0 & 1 \end{bmatrix} .$$

With the help of equations (4.20) and (4.21) we compute the results and tabulate for various steps of the single term PCBF technique in Table 4.1.

k	Initial state $\underline{x}(k-1)$	$\underline{X}^{(k-1)}$ Initial function	$\underline{P}^{(k)}$ Input	$\underline{V}^{(k)}$ Eq. (4.20)	$\underline{X}^{(k)}$ Eq. (4.21a)	$\underline{x}(k)$ Eq. (4.21b)
1	$\begin{bmatrix} 0 \\ 0 \end{bmatrix}$	$\begin{bmatrix} 0 \\ 0 \end{bmatrix}$	$\begin{bmatrix} 0 \\ 0 \end{bmatrix}$	$\begin{bmatrix} 0 \\ 0 \end{bmatrix}$	$\begin{bmatrix} 0 \\ 0 \end{bmatrix}$	$\begin{bmatrix} 0 \\ 0 \end{bmatrix}$
2	$\begin{bmatrix} 0 \\ 0 \end{bmatrix}$	$\begin{bmatrix} 0 \\ 0 \end{bmatrix}$	$\begin{bmatrix} 0 \\ 1.5 \end{bmatrix}$	$\begin{bmatrix} 0.575 \\ 1.15 \end{bmatrix}$	$\begin{bmatrix} 0.2875 \\ 0.575 \end{bmatrix}$	$\begin{bmatrix} 6.575 \\ 1.15 \end{bmatrix}$
3	$\begin{bmatrix} 0.575 \\ 1.15 \end{bmatrix}$	$\begin{bmatrix} 0.2817 \\ 0.575 \end{bmatrix}$	$\begin{bmatrix} 0 \\ 1.45 \end{bmatrix}$	$\begin{bmatrix} 1.544375 \\ 0.78875 \end{bmatrix}$	$\begin{bmatrix} 1.3471875 \\ 1.544375 \end{bmatrix}$	$\begin{bmatrix} 2.119315 \\ 1.93875 \end{bmatrix}$
4	$\begin{bmatrix} 2.119375 \\ 1.93875 \end{bmatrix}$	$\begin{bmatrix} 1.3471875 \\ 1.544375 \end{bmatrix}$	$\begin{bmatrix} 0 \\ 1.75 \end{bmatrix}$	$\begin{bmatrix} 1.8445417 \\ -0.1884165 \end{bmatrix}$	$\begin{bmatrix} 3.042083 \\ 1.8445418 \end{bmatrix}$	

Table 4.1. Single term PCBF solution of equation (4.24)

Table 4.2. gives the solution with m=2. The results by certain other methods are also shown for the sake of comparison.

k	t	Solution by single term PCBF		$x(k)$ by	
		$\underline{X}^{(k)}$	$\underline{x}(k)$	z Transform.	Fourth order Runge Kutta
0	0.0	0.0	0.0	0.0	0.0
1	0.5	0.0	0.0	0.0	0.0
2	1.0	0.067	0.0	0.0	0.0
3	1.5	0.345	0.134	0.131	0.173
4	2.0	0.912	0.556	0.550	0.630
5	2.5	1.726	1.267	1.260	1.370
6	3.0	2.680	2.186	2.180	2.273
7	3.5	3.634	3.174	3.170	3.137
8	4.0	4.470	4.094	4.100	4.013
9	4.5	5.122	4.846	4.860	4.752
10	5.0	5.587	5.398	5.410	5.287
11	5.5	5.917	5.778	5.780	5.675
12	6.0		6.058	6.050	5.984

Table 4.2. Solution of Equation (4.24) with m=2

Example 4.2. Consider the multiple delay system [D3].

$$\begin{aligned} \dot{x}(t) &= x(t-0.3) + x(t-0.2) - x(t-0.1) , \\ x(t) &= 1.0, \quad -0.3 \leq t < 0 , \\ x(0) &= 0.0 \ . \end{aligned} \tag{4.25}$$

Following Section 4.3 in the present case where $\underline{A}=0$, $\underline{L}_1=1.0$, $\underline{L}_2=1.0$, $\underline{L}_3=-1.0$, $\tau_1=0.3$, $\tau_2=0.2$, $\tau_3=0.1$, we take m=10. Then $N_1=3$, $N_2=2$, $N_3=1$. The single term PCBF solution may be computed from

$$\left.\begin{aligned} \underline{V}^{(k)} &= 0.1 [\underline{X}^{(k-3)} + \underline{X}^{(k-2)} - \underline{X}^{(k-1)}] \\ \underline{X}^{(k)} &= \tfrac{1}{2}\underline{V}^{(k)} + \underline{x}(k-1) \\ \underline{x}(k) &= \underline{V}^{(k)} + \underline{x}(k-1) \quad k=1,2,3,\ldots \end{aligned}\right\} . \tag{4.26}$$

Table 4.3 shows the solution and compares it with the results with other methods.

k	x(k) by Single term PCBF method	x(k) by Fourth order Runge-Kutta method	x(k) by Exact solution
0	0.0	0.0	0.0
1	0.100	0.117	0.100
2	0.295	0.292	0.295
3	0.380	0.362	0.380
4	0.371	0.362	0.371
5	0.387	0.379	0.387

Table 4.3. Solution of equation (4.25) with m=10

4.5. Stability of Computations

At the outset, let us consider the simple case of $\alpha=0$. The corresponding equation (4.13) may be rewritten as

$$\underline{x}(k) = \underline{\Psi}\,\underline{x}(k-1) + \underline{\zeta}\frac{\underline{L}}{m}\,\underline{x}^{(k-N)}\ , \tag{4.27}$$

where

$$\underline{\zeta} = (\underline{I} - \frac{\underline{A}}{2m})^{-1} \quad \text{and} \quad \underline{\Psi} = \underline{\zeta}(\underline{I} + \frac{\underline{A}}{2m}) \ .$$

Equations (4.14) and (4.15) imply that

$$\underline{x}^{(k-N)} = \frac{1}{2}\,[\underline{x}(k-N) + \underline{x}(k-N-1)] \ . \tag{4.28}$$

In view of these we write

$$\begin{aligned} \underline{x}(k) &= \underline{x}(k)\ , \\ \underline{x}(k+1) &= \underline{x}(k+1)\ , \\ \ldots & \quad \ldots \\ \underline{x}(k+N) &= \underline{\zeta}\frac{\underline{L}}{2m}[\underline{x}(k-1)+\underline{x}(k)] + \underline{\Psi}\,\underline{x}(k+N-1) \ . \end{aligned} \tag{4.29}$$

Defining

$$\underline{x}_k = [\underline{x}(k),\ \underline{x}(k+1),\ \ldots,\ \underline{x}(k+N)]^T\ ,$$

$$\underline{x}_{k-1} = [\underline{x}(k-1),\ \underline{x}(k), \ldots\ \underline{x}(k+N-1)]^T\ ,$$

equations (4.19) may be written in the form:

$$\underline{x}_k = \underline{G}_N\,\underline{x}_{k-1}\ , \tag{4.30}$$

where

$$\underline{G}_N = \left[\begin{array}{c|c|c|c} \underline{O}_{(nNxn)} & \multicolumn{3}{c}{\underline{I}_{(nNxnN)}} \\ \hline \underline{\zeta}\frac{\underline{L}}{2m} & \underline{\zeta}\frac{\underline{L}}{2n} & \underline{O}_{[nxn(N-2)]} & \underline{\Psi} \end{array}\right] \ . \tag{4.31}$$

Stability of computations can be assessed from the placement of the eigenvalues of $\underline{G}_N$ with respect to the unit circle.
If we consider the case when $0 \leq \alpha < 1$, it is possible to show that

$$\underline{G}_N = \left[\begin{array}{c|c|c|c|c} \multicolumn{1}{c|}{\underline{O}_{(nNxn)}} & \multicolumn{4}{c}{\underline{I}_{(nNxnN)}} \\ \hline (1-\alpha)\frac{L}{2m}\underline{\zeta} & \frac{L}{2m}\,\underline{\zeta} & \alpha\frac{L}{2m}\,\underline{\zeta} & \underline{O}_{[nxn(N-3)]} & \underline{\Psi} \end{array}\right] . \tag{4.37}$$

Stability of analysis of computations in the case of multiple delays may be carried out on similar lines. When the delay terms are dropped the above analysis reveals the stability of computations in lag-free problems.

V

SOLUTION OF FUNCTIONAL DIFFERENTIAL EQUATIONS

In this chapter we consider a class of functional equations in which certain terms have a stretched argument. Such equations arise out of the dynamics of the current collection system for an electric locomotive [D11]. We will employ the operational matrices for stretch in our treatment.

5.1. Solution of a vector differential equation containing a term with a stretched argument

Consider the equation

$$\begin{aligned} \dot{\underline{x}}(t) &= \underline{A}\ \underline{x}(t/\lambda) + \underline{B}\ \underline{x}(t), \\ \underline{x}(0) &= \underline{x}_0 , \end{aligned} \tag{5.1}$$

where $\underline{x}$ is a n-vector and $\underline{A}$ and $\underline{B}$ are matrices of appropriate dimensions. We first expand the various terms in (5.1) in m-size PCBF series as follows:

$$\left.\begin{aligned} &\dot{\underline{x}}(t) = \underline{V}\ \underline{\Theta}(t), \quad \underline{x}(t) = \underline{X}\ \underline{\Theta}(t) \\ &\underline{x}(t/\lambda) = \underline{\tilde{X}}\ \underline{\tilde{\Theta}}(t/\lambda) \text{ and} \\ &\underline{x}_0 = \underline{\hat{X}}\ \Theta(t) . \end{aligned}\right\} . \tag{5.2}$$

In the light of section 2.4.7we will write

$$\underline{x}(t/\lambda) = \underline{X}\ \underline{S}\ \underline{\Theta}(t) , \tag{5.3}$$

where $\underline{S}$ is the operational matrix for stretch. Using the operational matrix $\underline{E}$ for integration

$$\underline{x}(t) = \underline{V}\ \underline{E}\ \underline{\Theta}(t) + \underline{\hat{X}}\ \underline{\Theta}(t) . \tag{5.4}$$

Inserting (5.2-5.4) in (5.1)

$$\underline{V}\ \underline{\Theta}(t) = \underline{A}[\underline{V}\ \underline{S}\ \underline{E}\ \underline{\Theta}(t) + \underline{\hat{X}}\ \underline{\Theta}(t)] + \underline{B}[\underline{V}\ \underline{E}\ \underline{\Theta}(t) + \underline{\hat{X}}\ \underline{\Theta}(t)], \tag{5.5}$$

which implies that

$$\underline{V} = \underline{A}\ \underline{V}\ \underline{S}\ \underline{E} + \underline{B}\ \underline{V}\ \underline{E} + \underline{P}, \tag{5.6a}$$

where

$$\underline{P} = \text{PCBF spectrum of } \underline{A}\ \underline{x}_o + \underline{B}\ \underline{x}_o \text{ arranged as a matrix.} \tag{5.6b}$$

Equation (5.6) may be solved by the Kronecker product formula

$$\mathcal{V} = [\underline{I} + \underline{A} \otimes (\underline{S}\ \underline{E})^T - \underline{B} \otimes \underline{E}^T]^{-1} \mathcal{P}, \tag{5.7}$$

where $\mathcal{V}$ and $\mathcal{P}$ are long column vectors built from the columns of $\underline{V}$ and $\underline{P}$ respectively. We get x(t) from (5.4)

5.2. Single term PCBF technique

While stretching a signal in the interval [j-1,j), a portion of the signal goes to [j,j+1). This carry-over spectrum, obtained by an advance operation, is given by

$$\underline{X}(\underline{D}^{m-\alpha})^T\ \underline{\Theta}(t), \tag{5.8}$$

in which $\underline{D}$ corresponds to a delay matrix and α is the effective delay, τ_s, in terms of the number of subintervals. In each interval the advance (delay matrix transposed) and $\underline{S}$ will vary. With the carry-over equation (5.6b) becomes

$$\underline{P}^{(j)} = [\text{PCBF spectrum of } (\underline{A}\underline{x}_o + \underline{B}\underline{x}_o)] + [\underline{X}(\underline{D}^{m-\alpha})^T]^{(j-1)}.$$

We set m=1, maintaining the absolute size of the sub-interval. That is, we change the scale such that $\hat{t}$ = mt. The PCBF expansions and the resulting solutions are as follows:

$$\left.\begin{aligned} \dot{\underline{x}}(t) &= \underline{V}_s\ \underline{\Theta}\ (\hat{t}), \\ \underline{V}_s^{(j)} &= [\underline{I} - \frac{\underline{A}}{2m}\ \underline{S}^{(j)}_{(nx1)} - \frac{\underline{B}}{2m}]^{-1}\ \underline{\xi}^{(j)} \\ \dot{\underline{x}}(\hat{t}/\lambda) &= \underline{V}_s\ \underline{S}\ \underline{\Theta}\ (\hat{t})\ , \end{aligned}\right\}$$

$$\left.\begin{aligned} &\underline{X}^{(j)} = \frac{1}{2}\,\underline{V}_s^{(j)} + \underline{x}(j), \\ &\underline{x}(\hat{t}) = \underline{\chi}\,\underline{\Theta}\,(\hat{t}), \\ &\underline{x}(j+1) = \underline{x}(j) + \underline{V}_s^{(j)}, \\ &\text{where } \underline{\xi}^{(j)} = \frac{\underline{A}}{m}[\underline{P}^{(j)} + \underline{x}(j)] + \frac{\underline{B}}{m}\,\underline{x}(j). \end{aligned}\right\} \tag{5.9}$$

5.3. Examples

Example 5.1. Consider the equation of Fox et al [D5]

$$\begin{aligned} &\dot{x}(t) = x(0.8t) - x(t) \ , \\ &x(0) = 1.0 \ . \end{aligned} \tag{5.10}$$

Let $t=0.8\hat{t}$ and expand the terms in (5.10) in an 8-term PCBF series. We get the rate vector

$$\underline{V} = [\underline{I}+0.8\underline{E}\ \underline{S}+0.8\underline{E}]^{-1}\ (-0.8\hat{\underline{X}}_o\underline{S}-0.8\hat{\underline{X}}_o)\ , \tag{5.11}$$

where $\hat{X}_o$ is the PCBF expansion of the initial condition. Here $\lambda=1.25$. The results calculated through Walsh functions are shown in Fig. 5.1 comparing with those of Fox et al [D5]. The single term PCBF approach gives

$$\underline{V}^{(j)} = [1+\frac{1}{2m}(\underline{S}^{(j)}+1)]^{-1}\{-\frac{1}{m}\underline{x}(\hat{t}_{j-1})\underline{S}^{(j-1)}-\frac{1}{m}\underline{P}^{(j-1)}\}\ , \tag{5.12}$$

where

$$\underline{P}^{(j)} = \sum_{i=1}^{j-1} \underline{X}_i(\underline{S}_B)_{j+1,i};\quad \underline{P}^{(0)}=0,\ \underline{S}^{(j)} = (\underline{S}_B)_{j,j}\ .$$

Table 5.1 shows the results with m=10 comparing with the results of Fox et al [D5]. Recall that $\underline{S}_B$ is as given (2.36).

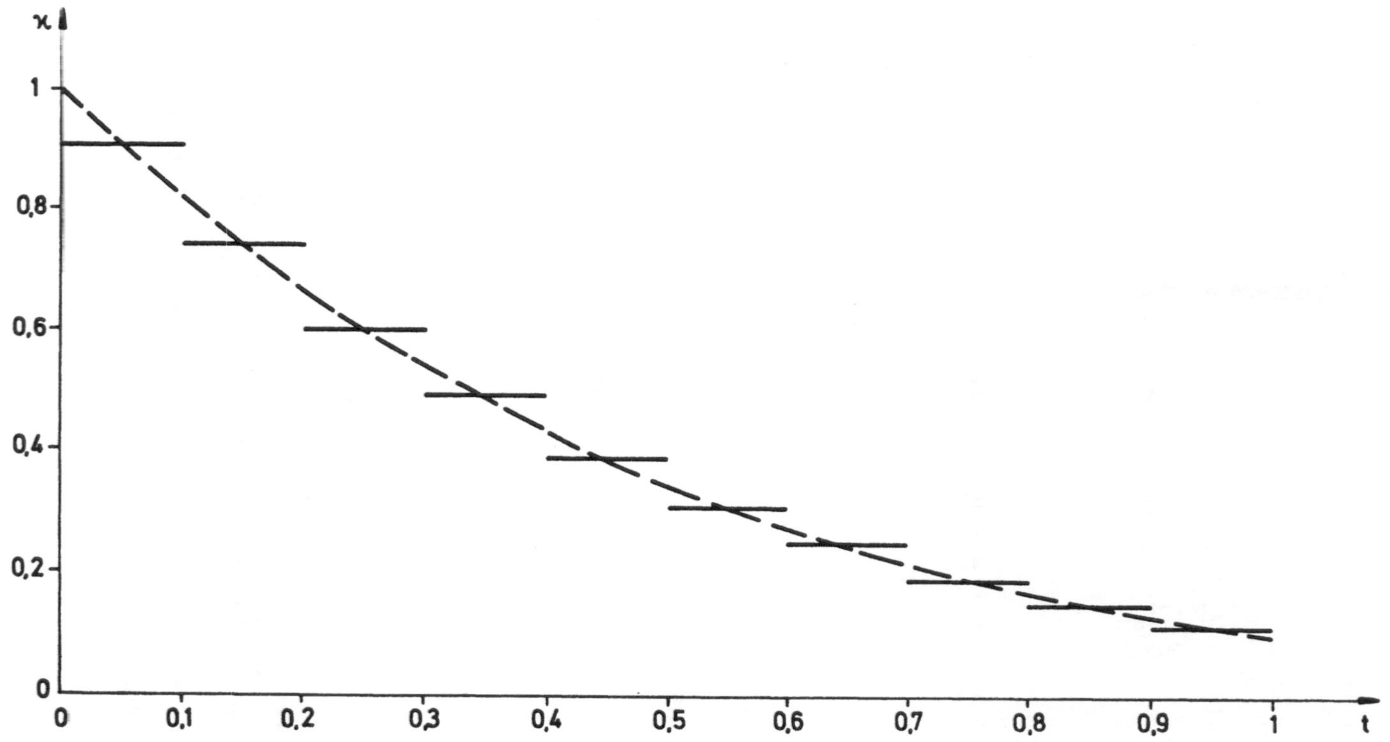

Fig. 5.1. Solution of example 5.1 (m=10). — — — Method of Fox et al [D5]. ——— PCBF solution

t	Fox et al [D5] Improved Finite Difference Method x(t)	Fox et al [D5] Deferred Correction Method x(t)	Single Term PCBF Method x(t)
0	1.0	1.0	1.0
0.1	0.816636	0.817054	0.818187
0.2	0.663989	0.664695	0.665621
0.3	0.537317	0.538210	0.538240
0.4	0.432562	0.433565	0.432426
0.5	0.346255	0.347307	0.3450201
0.6	0.275434	0.276486	0.275140
0.7	0.217572	0.218592	0.217532
0.8	0.170520	0.171487	0.170320
0.9	0.132459	0.133360	0.131885
1.0	0.101849	0.102673	0.100856

Table 5.1. Solution of Example 5.1.

Example 5.2.

Consider the equation [D10]

$$\dot{x}(t) = -x(t/\lambda), \quad x(0) = 1.0 \quad . \tag{5.13}$$

For m=10 and λ=2, 1.5, Fig. 5.2 shows the results by the single term PCBF method comparing well with those by a different method [D10].

Example 5.3.

Consider the equation [D5]

$$\begin{aligned} &\dot{x}(t) = -x(0.99t) - 0.95x(t) , \\ &x(0) = 1.0 \quad . \end{aligned} \tag{5.14}$$

Solution of (5.14) with m=10 is shown in Fig. 5.3.

Simple recurrence formulae in terms of BPF have been derived by Prasada Rao and Srinivasan [W54] for equations (5.10) and (5.14). Recalling the development in Section 2.4.7, which describes stretch operational matrices for m=J Ω, where $\Omega=\lambda/(\lambda-1)$, the single term PCBF solutions for equation (5.1) may be written as follows.

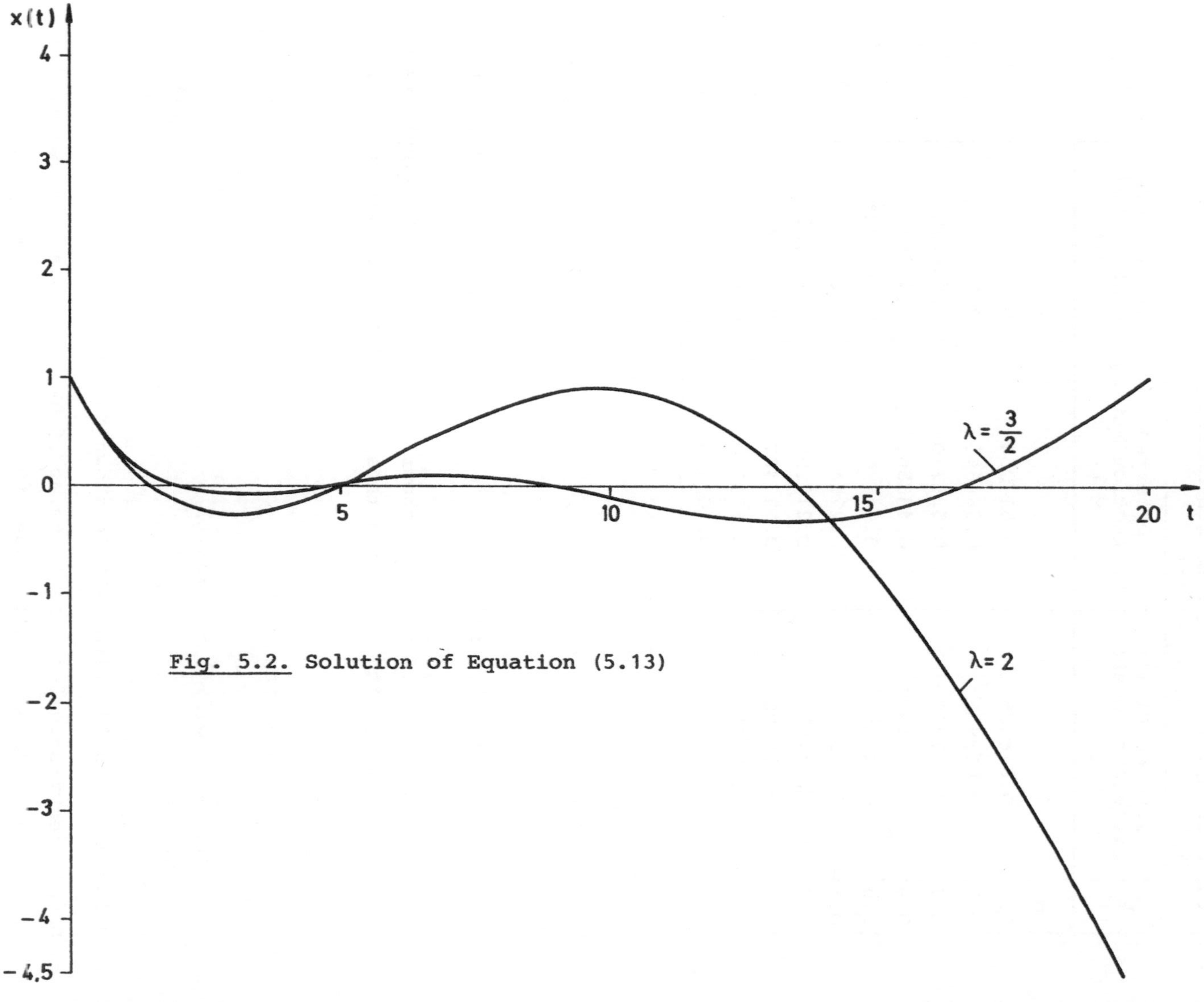

Fig. 5.2. Solution of Equation (5.13)

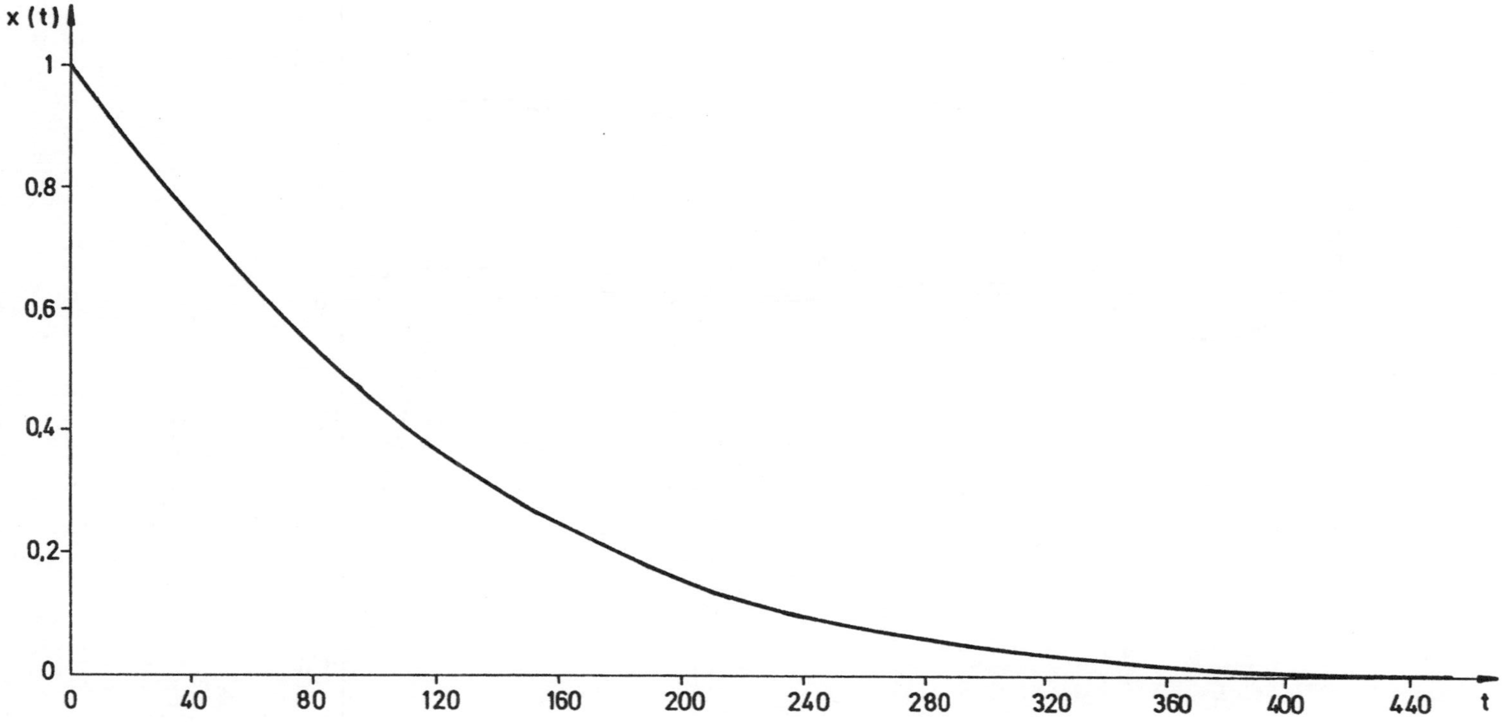

<u>Fig. 5.3.a.</u> Solution of example 5.3.
Subinterval ≐ 0.1 sec.

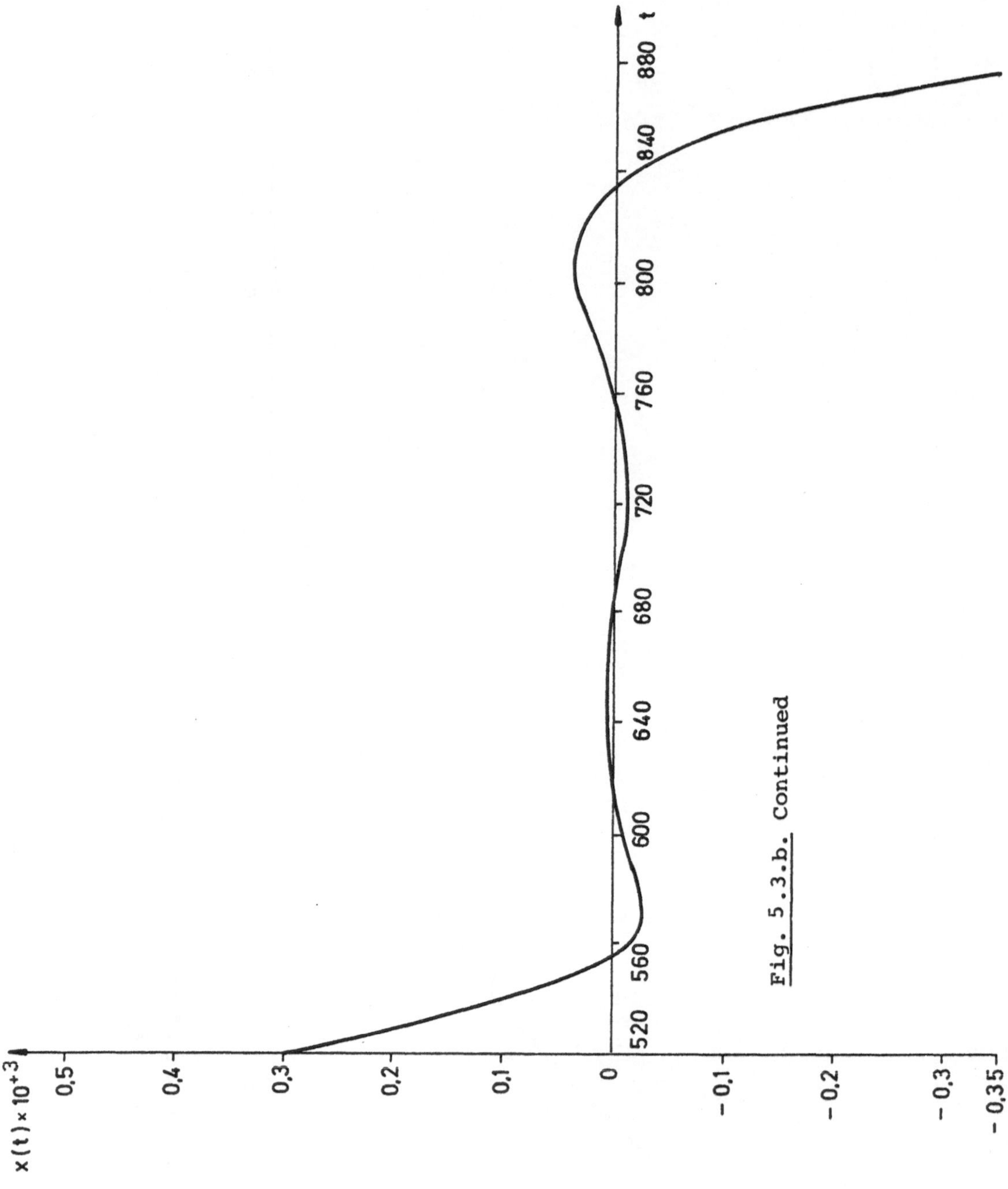

Fig. 5.3.b. Continued

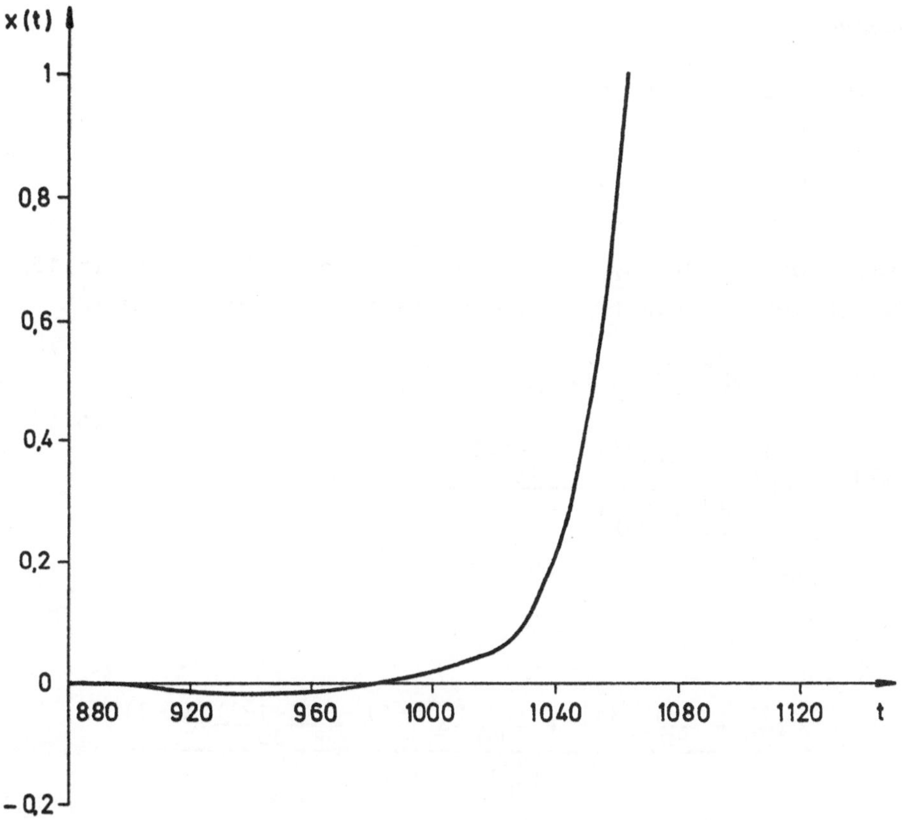

Fig. 5.3.c. Continued

Case 1: J=1

$$\underline{X}^{(i+1)} = [\underline{I} - \frac{\underline{B}}{2m} - \frac{\underline{A}}{2m}\frac{(\Omega-i-1)}{(\Omega-1)}]^{-1}$$

$$\cdot [(\underline{I} + \frac{\underline{B}}{2m} + \frac{\underline{A}}{2m}\frac{\Omega}{\Omega-1})\underline{X}^{(i)} + \frac{\underline{A}}{2m}\frac{(i-1)}{2m(\Omega-1)}\underline{X}^{(i-1)}] , \quad (5.15)$$

$i=1,2,\ldots, \Omega-1$.

Case 2: J>1

For $i=1$ to $\Omega-1$, we still use (5.15) and then use

$$\underline{X}^{(i+1)} = [\underline{I} - \frac{\underline{B}}{2m}]^{-1} [(\underline{I} + \frac{\underline{B}}{2m})\underline{X}^{(i)} + \frac{\underline{A}}{2m}\frac{1}{(\Omega-1)} \cdot$$

$$((\Omega-j-1)\underline{X}^{(i-k-1)} + (\Omega+1)\underline{X}^{(i-k)} + (j-1-l)\underline{X}^{(i-k-1)})],$$

for all $i=\Omega,\Omega+1,\ldots,m-1$. (5.16)

In the above,

k = the integer part of i/Ω,

$j = i-\Omega\cdot k$,

$$l = \left.\begin{matrix} -1 \\ 0 \end{matrix}\right\} \begin{matrix} \text{if } j=0 \\ \text{otherwise .} \end{matrix}$$

The solution of (5.10) may be written as a special case of (5.15), if $\Omega=5$ which corresponds to single segment expulsion. If $m=10$,

$$\underline{x}^{(i)} = [1 - \frac{(-2)}{20}]^{-1}\ 1 \quad \text{and}$$

$$\underline{x}^{(i+1)} = \frac{(\frac{19}{20} - \frac{1}{16})\underline{x}^{(i)} - \frac{i-1}{80}\underline{x}^{(i-1)}}{(\frac{21}{20} + \frac{4-i}{80})} \tag{5.17a}$$

$i=1,2,3$ and 4

and

$$\underline{x}^{(i+1)} = \frac{(\frac{19}{20} - \frac{4-j}{80})\underline{x}^{(i)} + \frac{(5+1)}{80}\underline{x}^{(i-1)} - \frac{(j-1-1)}{80}\underline{x}^{(i-1)}}{(\frac{21}{20})} \tag{5.17b}$$

$i=5,6,7,8$ and 9 ,

where $j=i-5$; $l=\begin{cases} -1 \text{ if } j=0 \\ 0 \text{ otherwise .} \end{cases}$

Let us now turn our attention to equation (5.14). We wish to compute the solution upto, say, $t=150$sec. We first let $\hat{t}=t/150$. Then,

$$\left.\begin{aligned} \frac{dx(\hat{t})}{d\hat{t}} &= -150x(0.99\hat{t}) + 142.5x(\hat{t}) \\ y(\hat{t}=0) &= 1 \end{aligned}\right\} . \tag{5.14a}$$

In our general formula in this case

$\underline{A} = -150$ and $\underline{B} = 142.5$.

For single segment expulsion, $\Omega=100$. (All vectors become scalars here)
Case 1: $J=1$, $m=\Omega$. Each segment is 1.5 sec.

$$\underline{x}^{(i)} = [1 - \frac{1}{200}(-7.5)]^{-1}\ 1\ ,$$

and

$$\underline{x}^{(i+1)} = [(1 + \frac{142.5}{200} - \frac{25}{33})\underline{x}^{(i)} - \frac{i-1}{132}\underline{x}^{(i-1)}]/[\frac{207.5}{200} - \frac{i}{132}]\ ,$$

$i=1,2,\ldots,99$.

Case 2: To improve accuracy we choose $J=10$, making $m=1000$. The segment width is now 0.15sec. The solution is now as follows

$$\underline{x}^{(i)} = [1 - \frac{1}{2000}(-7.5)]^{-1}.1,$$

$$\underline{x}^{(i+1)} = \frac{[(1 + \frac{142.5}{2000} - \frac{5}{66})\underline{x}^{(i)} - \frac{i-1}{1320}\underline{x}^{(i-1)}]}{[\frac{2007.5}{2000} - \frac{i}{1320}]}\ , \qquad (5.18a)$$

$i=1,2,\ldots,99$,

and

$$\underline{x}^{(i+1)} = [(1 + \frac{142.5}{2000})\underline{x}^{(i)} - \frac{150}{2000}\frac{1}{99}\{(99-j)\underline{x}^{(i-k+1)} + (100+1)\underline{x}^{(i-k)} + (j-1-1)\underline{x}^{(i-k-1)}\}]/[1 - \frac{142.5}{2000}], \qquad (5.18b)$$

$i=100,101,\ldots,999$,

where

k = integer part of $i/100$,
$j = i - 100k$,

$$l = \begin{cases} -1 & \text{if } j=0 \\ 0 & \text{otherwise.} \end{cases}$$

The solutions computed from (5.17) and (5.18) are shown in Figs. 5.3 and 5.4 respectively.
Recurrence formulae (5.17) and (5.18) have been derived for those cases with the following stretch matrices:
For $\lambda=1.25$, with single block-pulse expulsion $\Omega=5$ and

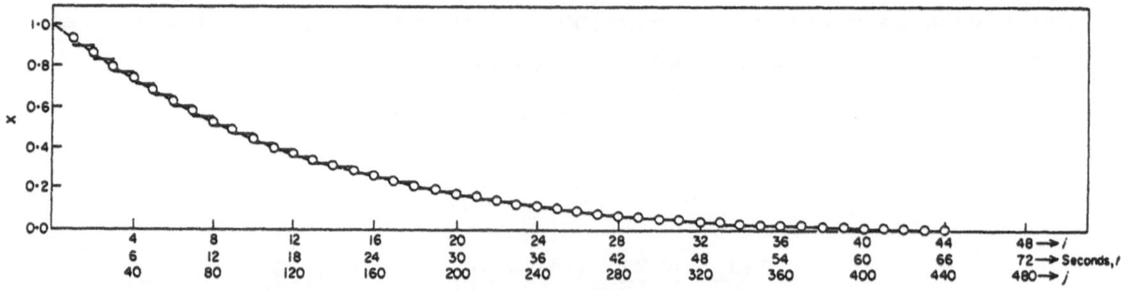

Fig.5.4a. Solution of dx(t)/dt=-x(0.99t)+0.95x(t), x(0)=1

----- Solution by Fox et al [D5]

— m=100 Solution by block pulse function method, over (0-78sec.)

o m=1000

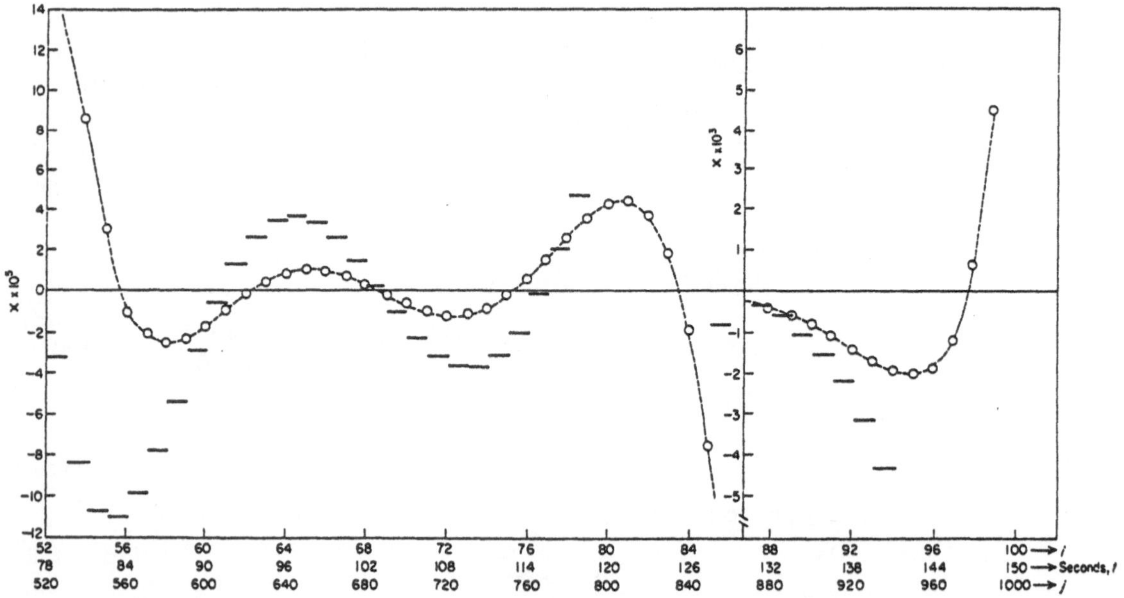

Fig.5.4b. Solution of dx(t)/dt=x(0.99t)+0.95x(t), x(0)=1

---- Solution by Fox et al [D5]

— m=100.Solution by block pulse function method over (78-150sec.)

o m=1000

$$\underline{S}_\beta = \begin{bmatrix} 4 & 1 & 0 & 0 & 0 \\ 0 & 3 & 2 & 0 & 0 \\ 0 & 0 & 2 & 3 & 0 \\ 0 & 0 & 0 & 1 & 4 \end{bmatrix} .$$

For m=10

$$\underline{S}_B = \left[\begin{array}{ccc} \underline{S}'_\beta & & \\ & & \underline{S}'_\beta \\ \hline 0 & \ldots & 0 \\ 0 & \ldots & 0 \end{array}\right] .$$

Next, in Example 5.3, since $\lambda=100/99$, for single block pulse expulsion $\Omega=100$ leading to

$$\underline{S}'_\beta = \begin{bmatrix} 99 & 1 & & & 0 \\ & 98 & 2 & & \\ & & \ddots & \ddots & 99 \\ & 0 & & & 1 \end{bmatrix} ,$$

and

$$\underline{S}_B = \left[\begin{array}{ccc} \underline{S}'_\beta & & 0 \\ & \underline{S}'_\beta \ddots & \\ 0 & & \underline{S}'_\beta \\ \hline & 0 & \end{array}\right] \begin{array}{l} \left.\vphantom{\begin{array}{c} a \\ b \\ c \end{array}}\right\} 990 \text{ rows} \\ \left.\vphantom{a}\right\} 10 \text{ rows} \end{array} .$$

VI

ANALYSIS OF NON-LINEAR AND TIME-VARYING SYSTEMS

Nonlinear and time varying dynamic systems are characterized in general by

$$\dot{\underline{x}} = \underline{f}(\underline{x},\underline{u},t), \ \underline{x}(0) = \underline{x}_o \ . \tag{6.1}$$

If we consider a time segment $\frac{1}{m}$ normalised to a unit interval and apply the ideas of single term PCBF expansions, we will be able to get an algebraic nonlinear relation of the form (for the i-th interval) as

$$\underline{V}^{(i)} = \underline{f}(\underline{X}^{(i)},\underline{U}^{(i)},t^{(i)}), \ i=1,2,\dots \ . \tag{6.2}$$

In view of the fact that in such a case $E = \frac{1}{2m}$,

$$\begin{aligned}\underline{X}^{(i)} &= \frac{1}{2m}\underline{f}(\underline{X}^{(i)},\underline{U}^{(i)},t^{(i)}) + \underline{x}_o \ , \\ \underline{X}^{(i+1)} &= \underline{X}^{(i)} + \frac{1}{2m}[\underline{f}(\underline{X}^{(i+1)},\underline{U}^{(i+1)},t^{(i+1)}) + \underline{f}(\underline{X}^{(i)},\underline{U}^{(i)},t^{(i)})], \\ & i \geq 1\end{aligned} \tag{6.3}$$

where $\underline{X}^{(i)}$, $\underline{U}^{(i)}$ and $\underline{t}^{(i)}$ are the PCBF coefficients of $\underline{x}(t)$, $\underline{u}(t)$ and t respectively over the subinterval $(\frac{i-1}{m},\frac{i}{m})$. Equation (6.3) is an implicit formula which requires an iterative predictor corrector type of algorithm for its solution [W57].

6.1. Solution of linear time varying systems

Consider a linear time-varying system

$$\dot{\underline{x}}(t) = \underline{A}(t)\underline{x}(t) + \underline{B}(t)\underline{u}(t), \ \underline{x}(0) = \underline{x}_o \ . \tag{6.4}$$

Solution of this equation by m-term Walsh series expansions and the corresponding BPF expansions used to be, before the extension technique and the single term WF method [W45] have appeared, by the usual method in which multiplication operators as discussed in chapter II were applied. The resulting algorithms is both algebraically and numerically quite tedious. With the advent of the single term technique (which may be seen to give formulae identical to the BPF method), the situation is very much simplified. We will straightaway write down the recursive formula with $t \in [0,\frac{1}{m})$ normalised to $\hat{t} \in [0,1)$. Let $\underline{V}^{(i)}$, $\underline{A}^{(i)}$, $\underline{X}^{(i)}$ and $\underline{P}^{(i)}$ be the PCBF coefficients of $\dot{\underline{x}}(t)$, $\underline{A}(t)$, $\underline{x}(t)$ and $\underline{B}(t)\underline{u}(t)$ respectively on the i-th sub-interval. The general formula (6.3) takes the following particular form:

$$\begin{aligned} \underline{X}^{(i)} &= \frac{1}{2m}[\underline{A}^{(i)}\underline{X}^{(i)}+\underline{P}^{(i)}], \\ \underline{X}^{(i+1)} &= \underline{X}^{(i)} + \frac{1}{2m}[\underline{A}^{(i+1)}\underline{X}^{(i+1)}+\underline{P}^{(i+1)}+\underline{A}^{(i)}\underline{X}^{(i)}\underline{P}^{(i)}], \\ & i=1,2,\ldots \; . \end{aligned} \tag{6.5}$$

Equation (6.5) may be written in an explicit form as

$$\left.\begin{aligned} \underline{X}^{(i)} &= \frac{\underline{P}^{(i)}}{2m}\left(\underline{I}-\frac{\underline{A}^{(i)}}{2m}\right)^{-1}, \\ \underline{X}^{(i+1)} &= [\underline{X}^{(i)}+\frac{1}{2m}(\underline{P}^{(i)}+\underline{P}^{(i+1)}+\underline{A}^{(i)}\underline{X}^{(i)})]\left(\underline{I}-\frac{\underline{A}^{(i+1)}}{2m}\right)^{-1}, \\ & i=1,2,\ldots \; . \end{aligned}\right\} \tag{6.6}$$

Example 6.1. Consider a linear time varying system

$$\dot{\underline{x}}(t) = \begin{bmatrix} 0 & 0 \\ t & 0 \end{bmatrix} \underline{x}(t), \qquad \underline{x}(0) = \begin{bmatrix} 1 \\ 1 \end{bmatrix} . \tag{6.7}$$

In this case $\underline{B}(t)\underline{u}(t) = 0$, $\underline{P}_i=0$ for all i. We expand the matrix $\underline{A}(t)$ over each interval in PCBF and apply the recurrence formula (6.6). The exact solution and the PCBF solution computed from (6.6) are shown in Fig.6.1. In the present computation m=4, and the exact solution is known to be

$$\begin{aligned} x_1(t) &= 1.0 , \\ x_2(t) &= 1 + \frac{1}{2}t^2 . \end{aligned}$$

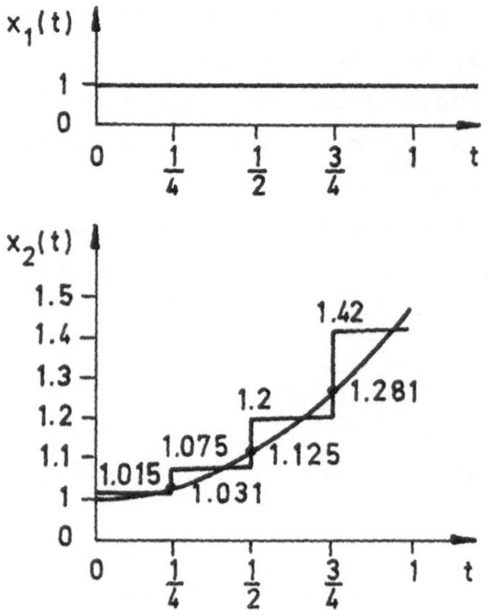

Fig. 6.1. Solution of equation (6.7)

Example 6.2. Consider the nonlinear time-delay system [D2].

$$\begin{aligned} \dot{x}(t) &= -x(t-1)-x^2(t),\ x(0)=0.1, \\ x(t) &= 0.1, \quad t\in[-1,0)\ . \end{aligned} \tag{6.8}$$

Changing the scale such that $\hat{t}=mt$,

$$\dot{x}(\hat{t}) = -\frac{1}{m}\,x(\hat{t}-m) - \frac{1}{m}x^2(\hat{t}).$$

The single term PCBF solution extended successively to contiguous sequence of subintervals may be shown to be (all vectors become scalars)

$$\underline{X}^{(k)} = 2\underline{X}^{(k)} - 2\underline{x}(k-1)\ , \tag{6.9a}$$

$$\frac{1}{m}\,[\underline{X}^{(k)}]^2 + 2\underline{X}^{(k)} + \frac{1}{m}\underline{X}^{(k-m)} - 2\underline{x}(k-1) = 0\ . \tag{6.9b}$$

With the root $X^{(k)}$ of (6.9b), the computed solution is shown in Fig. 6.2 from

$$\left.\begin{aligned} \underline{V}^{(k)} &= 2[\underline{X}^{(k)}-\underline{x}(k-1)]\ , \\ \underline{X}(k) &= \underline{x}(k-1) + \underline{V}^{(k)}. \end{aligned}\right\} \tag{6.10}$$

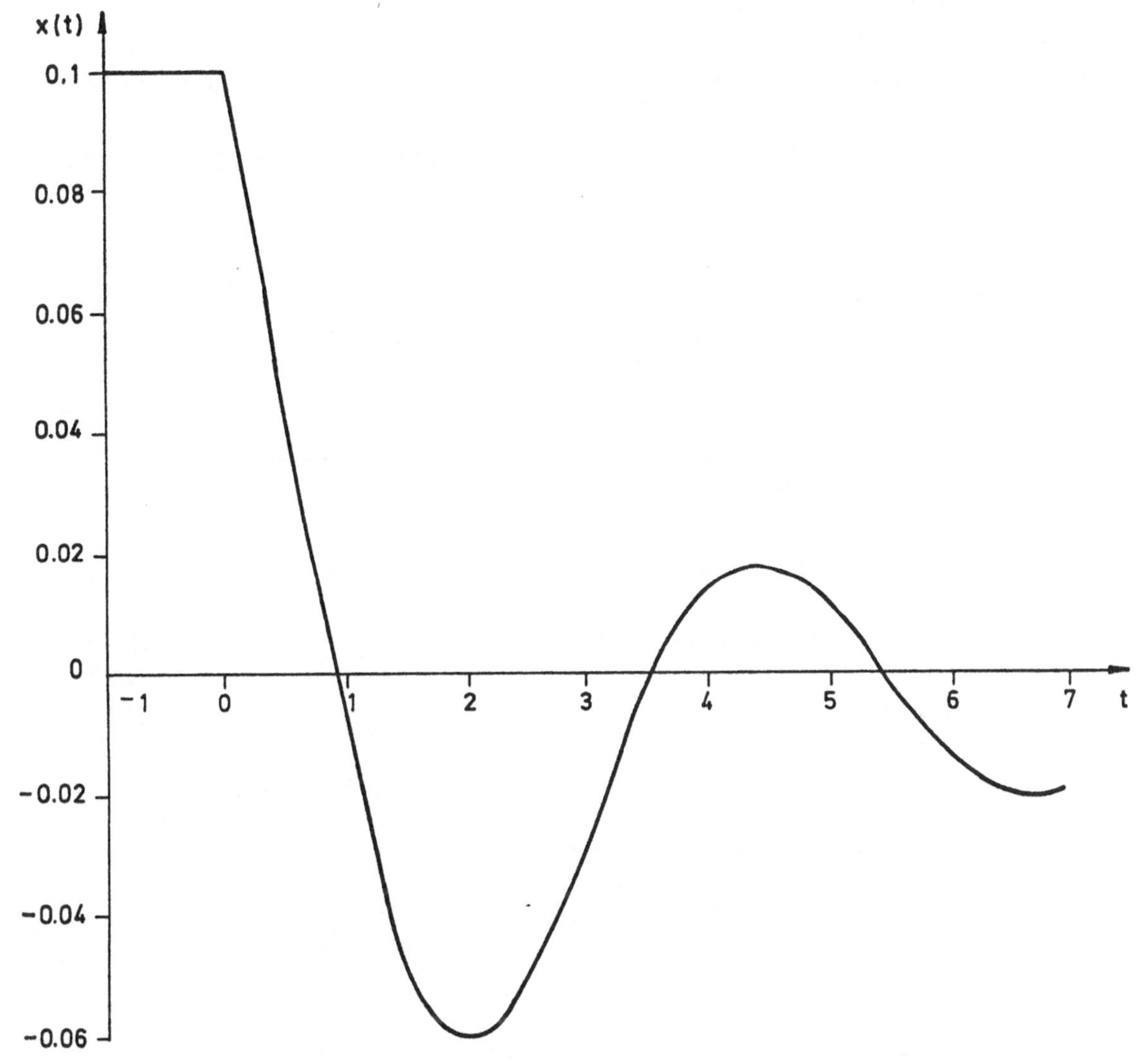

Fig. 6.2. Solution of equation (6.8)

Example 6.3. Consider the equation

$$\frac{dx}{dt} = -x - x^2 + 0.2, \quad x(0) = 0 \quad . \tag{6.11}$$

Let us compute the solution over segements of width, say, 0.1 sec. each. Change the time-scale such that [0,0.1) is normalised to [0,1). Then, (6.11) becomes

$$\frac{1}{0.1}\frac{dx}{d\hat{t}} = -x - x^2 + 0.2, \quad \hat{t} = \frac{t}{0.1} .$$

The solution may be obtained in a recursive form (all vectors are reduced to scalars here)

$$0.25[\underline{v}^{(i)}]^2 + \underline{v}^{(i)}[10.5+\underline{x}(j-1)] + [\underline{x}(j-1)]^2+\underline{x}(j-1)-0.2 = 0, \quad (6.12)$$

and

$$\left.\begin{aligned} \underline{x}^{(j)} &= \tfrac{1}{2}\underline{v}^{(j)} + \underline{x}(j-1) \\ \underline{x}(j) &= \underline{v}^{(j)} + \underline{x}(j-1) \end{aligned}\right\} \quad j=i=1,2,\ldots \quad . \qquad (6.13)$$

Fig. 6.3 shows this solution along with the one computed by Volterra series method [D8].

<u>Example 6.4.</u> Consider the well known Lienard equation [W53]

$$\ddot{x}(t) + \{x^2(t-\tau)-1\}\dot{x}(t-\tau)+x(t) = 0 \quad . \qquad (6.14)$$

Rewriting (6.14) as

$$\left.\begin{aligned} \dot{x}(t) &= x_2(t) \\ \dot{x}_2(t) &= -[x_1^2(t-\tau)-1]x_2(t-\tau)-x(t) \end{aligned}\right\}, \qquad (6.15)$$

The solution of (6.15) may be shown to be

$$\left.\begin{aligned} x1_1 &= \frac{1}{2m}x2_1 + x_1(0) , \\ x2_1 &= \frac{1}{2m}\{-(x1_{1-N}^2-1)x2_{i-N}-x1_1\} + x_2(0) , \\ &\text{and} \\ x1_{i+1} &= x1_i + \frac{1}{2m}[x2_{i+1}+x2_i] , \\ x2_{i+1} &= x2_i + \frac{1}{2m}[-(x1_{i+1-N}^2-1)x2_{i+1-N}-x1_{i+1} \\ &\quad -(x1_{i-N}^2-1)x2_{i-N}-x1_i] , \quad i=1,2,\ldots \end{aligned}\right\} \qquad (6.16)$$

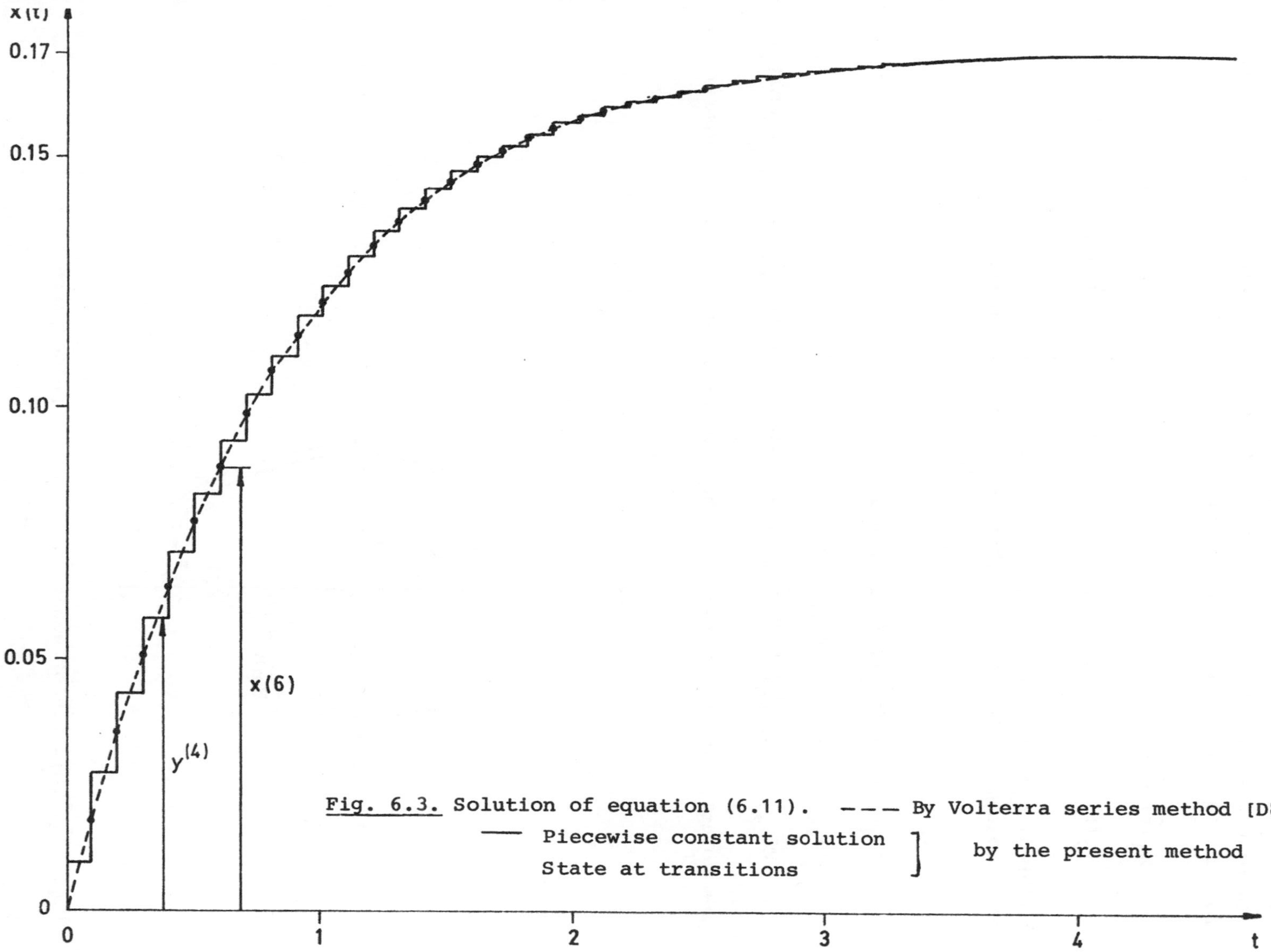

Fig. 6.3. Solution of equation (6.11). ––– By Volterra series method [D8]
— Piecewise constant solution, State at transitions } by the present method

where $X1$ and $X2$ represent the BPF values of x_1 and x_2 respectively. Consider the specific case with $\tau=0.15$, $x_1(0)=0$, $x_2(0)=2$, $x_1(t)=x_2(t)=0$ for $-\tau<t<0$. Let $t=10\hat{t}$. Then (6.15) becomes

$$\left.\begin{aligned} \dot{x}_1(\hat{t}) &= x_2(\hat{t}) , \\ \dot{x}_2(\hat{t}) &= -10[x_1^2(\hat{t}-0.75)-1]\;{}_2(\hat{t}-0.075) - 100x_1(\hat{t}) . \end{aligned}\right\} \qquad (6.17)$$

$x_1(0)=0$, $x_2(0)=20$, $x_1(\hat{t}) = x_2(\hat{t}) = 0$ for $-0.075<\hat{t}<0$.
We choose m=40 so that N=3, then (6.16) takes the special form

$$\begin{aligned} X1_1 &= \frac{1}{80} X2_1 , \\ X2_1 &= \frac{1}{80}[-(X1_{-2}^2-1)X2_{-2}-100X1_1]+20 , \\ X1_{i+1} &= X1_i + \frac{1}{80}[X2_{i+1}+X2_i] , \\ X2_{i+1} &= X2_i+\frac{1}{80}[-10(X1_{i-2}^2-1)X2_{i-2}-100X_{i+1}-10(X1_{i-3}^2-1)X2_{i-3}-100X1_i]. \end{aligned} \qquad (6.18)$$

The solution (6.18) is shown in phase plane in Fig. 6.4. It shows a

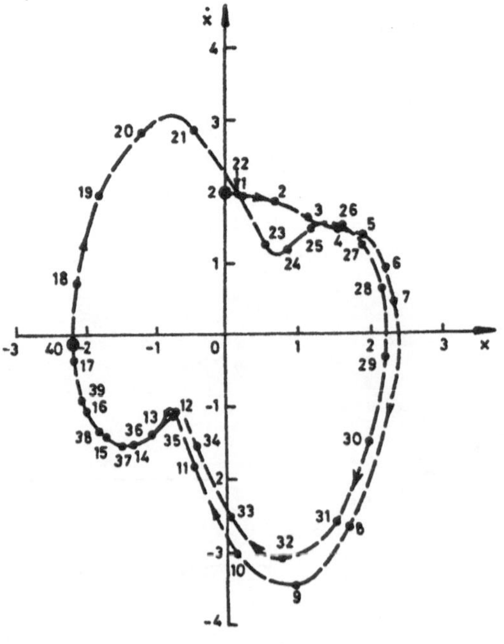

Fig. 6.4. Solution of equation (6.18) in phase plane

period of approximately 5.75 sec. and agrees well with that obtaines by Grafton [D6]. The PCBF solution in phase plane is a set of points. At each of these points a period of time equal to 1/m on the normalised scale is lumped. The dotted line through these points shows the phase trajectory [W53].

OPTIMAL CONTROL OF LINEAR LAG-FREE SYSTEMS

7.1. Time-invariant systems [W11, W14]

The optimal control of a time-invariant linear system

$$\dot{\underline{x}} = \underline{A}\underline{x} + \underline{B}\underline{u} , \tag{7.1}$$

with the quadratic performance index

$$J = \frac{1}{2} \int_{o}^{t_f} (\underline{x}^T \underline{Q}\underline{x} + \underline{u}^T \underline{R}\underline{u}) dt , \tag{7.2}$$

is well known to be

$$\underline{u}^* = \underline{R}^{-1}\underline{B}^T\underline{p}(t) , \tag{7.3}$$

where p(t) satisfies the following equation

$$\left.\begin{array}{l} \begin{bmatrix} \dot{\underline{x}} \\ \dot{\underline{p}} \end{bmatrix} = \begin{bmatrix} \underline{A} & \underline{B}\underline{R}^{-1}\underline{B}^T \\ \underline{O} & -\underline{A}^T \end{bmatrix} \begin{bmatrix} \underline{x} \\ \underline{p} \end{bmatrix} \\ \underline{p}(t_f) = 0 \\ \underline{x}(0) = \underline{x}_o \end{array}\right\} . \tag{7.4}$$

It is convenient to write (7.4) by letting $\hat{t} = t_f - t$ as

$$\begin{bmatrix} \dot{\underline{x}}(\hat{t}) \\ \dot{\underline{p}}(\hat{t}) \end{bmatrix} = \begin{bmatrix} -\underline{A} & -\underline{B}\underline{R}^{-1}\underline{B}^T \\ -\underline{Q} & \underline{A}^T \end{bmatrix} \begin{bmatrix} x(\hat{t}) \\ p(\hat{t}) \end{bmatrix} \triangleq \hat{A} \begin{bmatrix} \underline{x}(\hat{t}) \\ \underline{p}(\hat{t}) \end{bmatrix} . \tag{7.5}$$

The state transition matrix of (7.5) is

$$\exp(-\hat{\underline{A}}\,\hat{t}) = \begin{bmatrix} \underline{\eta}_{11}(\hat{t}) & \underline{\eta}_{12}(\hat{t}) \\ \underline{\eta}_{21}(\hat{t}) & \underline{\eta}_{22}(\hat{t}) \end{bmatrix} . \tag{7.6}$$

Since $\underline{p}(\hat{t}=0)=0$, the solution of (7.5) can be written as

$$\underline{x}(\hat{t}) = \underline{\eta}_{11}(\hat{t})\underline{x}(\hat{t}=0), \tag{7.7}$$

$$\underline{p}(\hat{t}) = \eta_{21}(\hat{t})\underline{x}(\hat{t}=0). \tag{7.8}$$

From (7.7)

$$\underline{x}(\hat{t}=0) = \underline{\eta}_{11}^{-1}(\hat{t})\underline{x}(\hat{t}).$$

Then,

$$\underline{p}(\hat{t}) = \underline{\eta}_{21}(\hat{t})\underline{\eta}_{11}^{-1}(\hat{t})\underline{x}(\hat{t}),$$

giving

$$\underline{u}^*(t) = \underline{R}^{-1}\underline{B}^T\underline{\eta}_{21}(t_f-t)\underline{\eta}_{11}^{-1}(t_f-t)\underline{x}(t_f-t)$$

$$\triangleq -\underline{K}(t_f-t)\underline{x}(t_f-t), \tag{7.9}$$

where $\underline{K}(\hat{t})$ is the optimal feedback gain matrix.

PCBF solution of the above problem gives piecewise constant gain elements which are convenient to implement in a practical situation. Without loss of generality, we first change the time scale for normalisation by letting

$$\bar{t} = \hat{t}/t_f \tag{7.10}$$

and write the state-costate vector as

$$\begin{bmatrix} \underline{x}(\bar{t}) \\ \underline{p}(\bar{t}) \end{bmatrix} \triangleq \underline{x}_p(\bar{t}) \quad . \tag{7.11}$$

Then (7.5) becomes

$$\dot{\underline{x}}_p = -t_f\,\underline{\hat{A}x}_p, \quad 0\leq\bar{t}\leq 1 \quad . \tag{7.12}$$

If we expand the state-costate-rate vector in m-size PCBF as

$$\underline{\dot{x}}_p(\bar{t}) = \underline{V}\ \underline{\theta}(\bar{t})$$

the state-costate equation reduces to

$$\underline{V} = \underline{\bar{A}}\ \underline{V}\ \underline{E} + \underline{P}\ ,\ \bar{A} \text{ corresponding to } \bar{t} \qquad (7.13)$$

Here P contains only the initial state-costate terms. Equation (7.13) is in the form of (3.5) for which we have obtained a recurrence formula using the single term PCBF technique with facility for extended computation. A general computer program written to handle (3.5) may be used to solve above optimal control problem.

Example 7.1.: Let us consider the problem

$$\underline{\dot{x}} = \begin{bmatrix} 0 & 0 \\ 1 & 0 \end{bmatrix} \underline{x} + \begin{bmatrix} 1 \\ 0 \end{bmatrix} \underline{u},\quad \underline{x}(0) = \begin{bmatrix} 0 \\ 10 \end{bmatrix} , \qquad (7.14)$$

$\underline{x}(t_f)$ unspecified,

$$J = \frac{1}{2} \int_0^{t_f} (\underline{x}^T \underline{Qx} + \underline{u}^T \underline{Ru})dt\ , \qquad (7.15)$$

where

$$\underline{Q} = \begin{bmatrix} 0 & 0 \\ 0 & 4 \end{bmatrix},\ \underline{R} = 1\ ,$$

$$t_f = \pi/2.$$

In this case

$$\underline{\hat{A}} = \begin{bmatrix} 0 & 0 & -\frac{\pi}{2} & 0 \\ -\frac{\pi}{2} & 0 & 0 & 0 \\ 0 & 0 & 0 & \frac{\pi}{2} \\ 0 & -\frac{\pi}{2} & 0 & 0 \end{bmatrix}.$$

The exact feedback gains are given by

$$k_1(t) = \frac{[\sinh(\pi-2t)-\sin(\pi-2t)]}{[\cosh^2(\frac{\pi}{2}-t)+\cos^2(\frac{\pi}{2}-t)]},$$

$$k_2(t) = \frac{[\cosh(\pi-2t)-\cos(\pi-2t)]}{[\cosh^2(\frac{\pi}{2}-t)+\cos^2(\frac{\pi}{2}-t)]}.$$

The PCBF based results along with the above are shown in Fig. 7.1.

7.2. Time varying systems [W18, W37]

Consider the system

$$\dot{\underline{x}}(t) = \underline{A}(t)\underline{x}(t) + \underline{B}(t)\underline{u}(t),\ \underline{x}(0) = \underline{x}_o, \qquad (7.16)$$

with the performance index

$$J = \frac{1}{2}\int_o^{t_f} (\underline{x}^T\underline{Q}(t)\underline{x} + \underline{u}^T\underline{R}(t)\underline{u})dt. \qquad (7.17)$$

The optimal control is well known to be

$$\underline{u}(t) = \underline{R}^{-1}(t)\underline{B}^T(t)\underline{p}(t), \qquad (7.18)$$

where the adjoint variable p(t), an n-vector, satisfies the canonical equation

$$\dot{\underline{x}}_p(t) = \hat{\underline{A}}(t)\underline{x}_p,\ \underline{p}(t_f) = 0,$$

$$\underline{x}_p = \begin{bmatrix}\underline{x}\\ \underline{p}\end{bmatrix},\quad \hat{\underline{A}} = \begin{bmatrix}\underline{A}(t) & \underline{B}(t)\underline{R}^{-1}(t)\underline{B}^T(t)\\ \underline{Q}(t) & -\underline{A}^T(t)\end{bmatrix}. \qquad (7.19)$$

At this stage, it should be noted that (7.19) is in the form of (6.4). The single term PCBF method outlined in section 6.1. is readily applicable. We choose a suitable sub-interval of duration $\frac{1}{m}$ and normalise the time scale to make this segment [0,1). All the other procedural steps such as changing of variable to $\hat{t}$ to render backward integration possible and taking the time-varying matrix $\hat{\underline{A}}$ as constant during the subinterval etc.,can be taken to compute the optimal feedback gains as in the case of the time invariant case.

In the process of continuing the solution over the successive segments, the PCBF approximation of the matrix $\underline{\hat{A}}$ should be appropriately employed at each step.

Example 7.2. [W18, W37]
Let a time-varying linear system described by

$$\dot{x}(t) = t\,x(t) + u(t), \quad x(0) = 1 \tag{7.20}$$

be optimized with respect to

$$J = \frac{1}{2}\int_0^1 (x^2+u^2)\,dt \tag{7.21}$$

In this case

$$\underline{\hat{A}}(t) = \begin{bmatrix} t & 1 \\ 1 & -t \end{bmatrix} .$$

Following the procedure outlined in section 7.2. the feedback gain $k(t)$ is calculated and shown in Fig. 7.2 along with the result obtained through the solution of Riccati equation. Over the normal interval $[0,1)$, m is taken as 4. The single term PCBF solution is extended over the four segments. One advantage of the single term PCBF approach is evidently the simplicity of the algorithm requiring no special operations for multiplication. Each sub-interval is efficiently decoupled from the rest and the product terms become simple. Over each sub-interval the system state-costate matrix $\underline{\hat{A}}(t)$ is treated as constant. The algorithm derived for time-invariant systems is only slightly modified to suit time-varying systems.

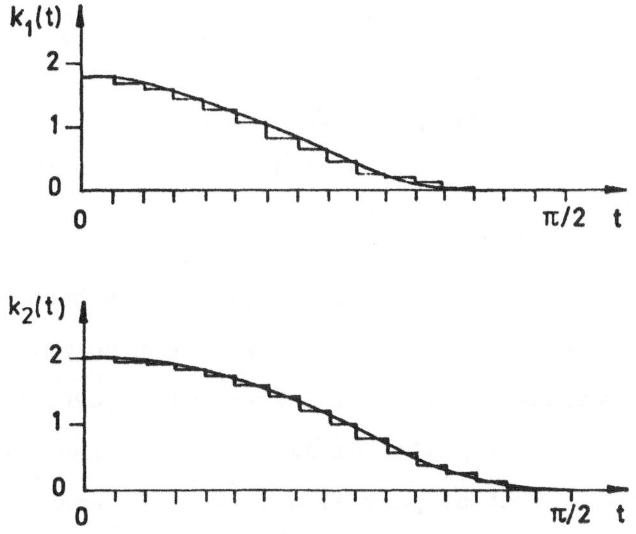

<u>Fig. 7.1.</u> Solution of example 7.1.

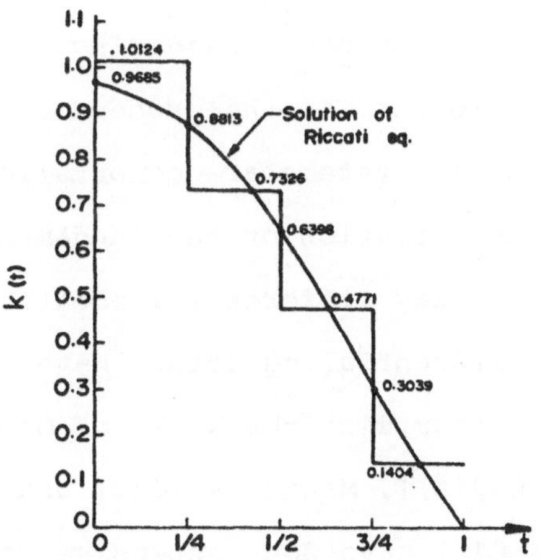

<u>Fig. 7.2.</u> Solution of example 7.2.

VIII

OPTIMAL CONTROL OF TIME-LAG SYSTEMS

[W36, W43(a)]

Optimal control of time delay systems has been of considerable interest to control engineers in recent years. The early works of Kharatishivili [C25] and Krasovskii [C26, C27] provided the basic foundations. Eller et al. [C19] use the basic formulation to develop a set of partial differential equations (PDE) to be solved for optimal feedback control of a linear TVP system with delay in its state alone. A parameter imbedding method of design has been suggested by Chan and Perkins [C12], Gracovetsky and Vidyasagar [C20] and Jamshidi and Malek-Zavarei [C24] for such problems. Inoue et al. [C22] gave a sensitivity technique and Sannuti [C44] suggested a singular perturbation approach to the case of systems with small delay. Ray and Soliman [C39] and Budelis and Bryson [C11] gave a PDE method considering delays both in the state and control. Major problems in the process of design of optimal controls for time-delay systems are computational in nature. Often we require either the solution or the fundamental matrix of the system described by delay differential equations. Several important aspects of delay differential equations have been recently discussed by Tsoi [C47-C50] summarizing the works of Olbrot [C33-C35], Popov [C37], Choudhury [C13(a)], Manitius [C32] and others. Banks and his colleagues [C4-C10] have done extensive work in hereditary systems using averaging approximation technique with function space analysis. Delfour [C14-C18] and colleagues gave rigorous convergence arguments for the state, costate and Riccati operator variables as well as for the optimal controls for the approximating discrete system control problems. The imbedding approach in this case leads simultaneously to differential equations with delayed and advanced arguments. We will employ PCBF to solve these equations and the related optimal control problems.

8.1. Optimal control of linear systems with state delay

Consider a system described by

$$\dot{\underline{x}}(t) = \underline{A}\ \underline{x}(t) + \underline{L}\ \underline{x}(t-\tau_s) + \underline{B}\ \underline{u}(t), \qquad (8.1a)$$

with initial data

$$\underline{x}(0) = \underline{x}_o\,,$$

$$\underline{x}(t) = \underline{x}_b(t),\ t \in [t_o - \tau_s,\ t_o],$$

where $\underline{x}$ is an n-vector of state, t_o initial time ($t_o = 0$ without loss of generality), $\underline{u}$ an r-vector of inputs and t_f represents final time. $\underline{A}$, $\underline{L}$ and $\underline{B}$ are matrices of appropriate dimensions. τ_s is the fixed time-delay in the plant and the initial function $\underline{x}_b(t)$ is continuous in its interval. The problem is to minimize

$$J = \frac{1}{2}\,\underline{x}^T(t_f)\ \underline{G}\ \underline{x}(t_f) + \frac{1}{2}\int_0^{t_f} (\underline{x}^T\ \underline{Q}\ \underline{x} + \underline{u}^T\ \underline{R}\ \underline{u})\ dt\,, \qquad (8.1b)$$

where $\underline{Q}$ and $\underline{R}$ are constant positive definite matrices. $\underline{R}$ is symmetric. We wish to find $\underline{u}(t)$, $0 \leq t \leq t_f$, which under the above conditions of fixed initial state, free final state and fixed terminal time, minimizes J subject to (8.1a). It is well known that the necessary and sufficient conditions can be expressed as the two point boundary value problem (TPBVP)

$$\dot{\underline{x}}(t) = \underline{A}\,\underline{x}(t) + \underline{L}\,\underline{x}(t-\tau_s) - \underline{B}\,\underline{R}^{-1}\,\underline{B}^T\,\underline{p}(t),$$

$$0 \leq t \leq t_f, \tag{8.2a}$$

$$\dot{\underline{p}}(t) = -\underline{Q}\,\underline{x}(t) - \underline{A}^T\,\underline{p}(t) - \underline{L}^T\,\underline{p}(t+\tau_s),$$

$$0 \leq t \leq t_f - \tau_s, \tag{8.2b}$$

$$= -\underline{Q}\,\underline{x}(t) - \underline{A}^T\,\underline{p}(t), \quad t_f - \tau_s \leq t \leq t_f,$$

$$\underline{p}(t_f) = 0. \tag{8.2c}$$

The optimal control

$$\underline{u}^*(t) = -\underline{R}^{-1}\,\underline{B}^T\,\underline{p}(t), \quad 0 \leq t \leq t_f. \tag{8.2d}$$

Let

$$\underline{p}(t) = \underline{K}(t)\,\underline{x}(t) + \underline{\tilde{h}}(t).$$

Riccati transformation of (8.2) yields

$$\left.\begin{aligned}\underline{\dot{K}}(t) &= -\underline{K}(t)\,\underline{A} - \underline{A}^T\,\underline{K}(t) + \underline{K}(t)\,\underline{B}\,\underline{R}^{-1}\,\underline{B}^T\,\underline{K}(t) - \underline{Q},\\ \underline{K}(t_f) &= \underline{G},\\ \underline{K}(t) &= 0,\ t > t_f,\end{aligned}\right\}\quad (8.3a)$$

$$\left.\begin{aligned}\underline{\dot{\tilde{h}}}_N(t) &= [\underline{K}(t)\,\underline{B}\,\underline{R}^{-1}\,\underline{B}^T - \underline{A}^T]\,\underline{\tilde{h}}_N - \underline{K}(t)\,\underline{L}\,\underline{x}_{N-1}(t-\tau_s)\\ &\quad -\underline{L}^T\,\underline{K}(t+\tau_s)\,\underline{x}_{N-1}(t+\tau_s)\\ &\quad -\underline{L}^T\,\underline{\tilde{h}}_{N-1}(t+\tau_s),\ \underline{\tilde{h}}(t) = 0,\ t \geq t_f\end{aligned}\right\}\quad (8.3b)$$

$$\left.\begin{aligned}\underline{\dot{x}}_N(t) &= [\underline{A} - \underline{B}\,\underline{R}^{-1}\,\underline{B}^T\,\underline{K}(t)]\,\underline{x}_N - \underline{B}\,\underline{R}^{-1}\,\underline{B}^T\,\underline{\tilde{h}}_N(t)\\ &\quad + L\,\underline{x}_{N-1}(t-\tau_s),\ \underline{x}(0) = \underline{x}_0,\\ \underline{x}(t) &= \underline{x}_b(t),\ t \in [t_0-\tau_s,\ t_0).\end{aligned}\right\}\quad (8.3c)$$

The original TPBVP with forcing terms is reduced to initial value problems with delay and advance arguments. (8.3a) is integrated backward in time to get $\underline{K}(t)$ just once. The adjoint equation (8.3b) is integrated backwards in time and (8.3c) is integrated forward in time iteratively. The stage index of iteration is denoted by N. The iterative solution is shown to converge to the optimal solution [W36].

8.2. Solution of Riccati equation by single term PCBF method

Equations (8.2a) and (8.2b) are written in the form

$$\begin{bmatrix} \dot{\underline{x}}(t) \\ \dot{\underline{p}}(t) \end{bmatrix} = \begin{bmatrix} \underline{A} & -\underline{B}\,\underline{R}^{-1}\,\underline{B}^T \\ -\underline{Q} & -\underline{A}^T \end{bmatrix} \begin{bmatrix} \underline{x}(t) \\ \underline{p}(t) \end{bmatrix} + \text{delay terms} .$$

We obtain the transient solution in the form

$$\begin{bmatrix} \dot{\underline{x}}(t) \\ \dot{\underline{p}}(t) \end{bmatrix} = -\hat{\underline{A}} \begin{bmatrix} \underline{x}(t) \\ \underline{p}(t) \end{bmatrix}, \tag{8.4}$$

where $\hat{\underline{A}}$ is as defined in chapter VII.

To begin the reduction to the single term PCBF form we let $\hat{t} = m.t$

$$\begin{aligned} \underline{V}_k^{(i)} &= \left[\underline{I} + \frac{\hat{\underline{A}}}{2m}\right]^{-1} \left[\frac{\hat{\underline{A}}}{2m}\,\underline{\rho}(i)\right], \\ \underline{K}_y^{(i)} &= \frac{1}{2}\,\underline{V}_k + \underline{\rho}(i), \\ \underline{p}(i-1) &= \underline{V}_k^{(i)} + \underline{\rho}(i),\ i = mt_f \ \ldots\ 3,2,1 \text{ and } \underline{\rho}(mt_f) = \begin{bmatrix} I \\ O \end{bmatrix}. \end{aligned} \tag{8.5}$$

In the above,

$$\underline{\eta}(t_f,t) = \underline{\rho}(i) \text{ and } \overline{\underline{\eta}}(t_f,t) = \underline{K}_y^{(i)},$$

$$\underline{\eta}(t_f,t) = \begin{bmatrix} \underline{\eta}_{11}(t_f,t) & \underline{\eta}_{12}(t_f,t) \\ \underline{\eta}_{21}(t_f,t) & \underline{\eta}_{22}(t_f,t) \end{bmatrix}.$$

Thus,

$$\begin{bmatrix} \underline{x}(t_f) \\ \underline{p}(t_f) \end{bmatrix} = \begin{bmatrix} \underline{n}_{11}(t_f,t) & \underline{n}_{12}(t_f,t) \\ \underline{n}_{21}(t_f,t) & \underline{n}_{22}(t_f,t) \end{bmatrix}.$$

Following Levine [C30],

$$\left.\begin{aligned} &\underline{K}(i) = \underline{n}_{22}^{-1}(t_f,t)\ \underline{n}_{21}(t_f,t), \\ &\text{and} \\ &\underline{Y}_k^{(i)} = \overline{\underline{n}}_2^{-1}(t_f,t_f)\ \overline{\underline{n}}_{21}(t_f,t). \end{aligned}\right\} \quad (8.6)$$

$\underline{K}(i)$ is the discrete solution of the Riccati equation.

8.3. Solution of adjoint and state equations with delay and advance arguments

Using the solution of the Riccati equation, we solve (8.3b) and (8.3c) iteratively employing the operational matrices for delay and advance in the single term PCBF approach. Equation (8.3b) has to be integrated backwards with $\tilde{h}(t) = 0$, $t \geq t_f$ and $\hat{t} = m.t$. We represent in the i-th interval the single term PCBF components of $\underline{K}(\hat{t})$, $\underline{K}(\hat{t}\pm\nu)$, $\underline{x}(\hat{t}\pm\nu)$, $\dot{\underline{x}}(\hat{t})$, $\underline{x}(\hat{t})$, $\dot{\tilde{\underline{h}}}(t)$, $\tilde{\underline{h}}(\hat{t})$, and $\tilde{\underline{h}}(\hat{t}\pm\nu)$ with known delay and advance ν as

$$\underline{K}(\hat{t}) = \underline{Y}_k^i\ \theta_1(\hat{t});\ \underline{K}(\hat{t}\pm\nu) = \underline{Y}_k^{i\pm N_d}\ \theta_1(\hat{t});$$

$$\dot{\underline{x}}(\hat{t}) = \underline{V}_k^i\ \theta_1(\hat{t});\ \underline{x}(\hat{t}) = \underline{Y}_x^i\ \theta_1(\hat{t});$$

$$\underline{x}(\hat{t}\pm\nu) = \underline{Y}_x^{i\pm N_d}\ \theta_1(\hat{t});$$

$$\dot{\underline{\tilde{h}}}(\hat{t}) = \underline{V}_h^i\ \theta_1(\hat{t});\ \underline{\tilde{h}}(t) = \underline{Y}_h^i\ \theta_1(t);$$

$$\underline{\tilde{h}}(\hat{t}\pm\nu) = \underline{Y}_h^{i\pm N_d}\ \theta_1(\dot{\hat{t}}),$$

where $N_d = \nu = m.\tau_s$.

Equation (8.3b) has to be integrated backwards with the single term PDBF as

$$\underline{V}_{h,N}^i = -\frac{1}{m}\{\underline{\xi}^i[\underline{V}_{h,N}^i\ \underline{E} + \underline{\tilde{h}}(i)] + \underline{G}_{N-1}^i\},$$

$$\underline{\xi}^i = (\underline{Y}_k^i\ \underline{B}\ \underline{R}^{-1}\ \underline{B}^T - \underline{A}^T),$$

$\underline{\tilde{h}}(i)$ = discrete time value of $\tilde{h}(t)$,

$$\underline{G}_{N-1}^i = \{-\underline{Y}_k^i(\underline{L}\ \underline{Y}_x^{i-N_d}) - \underline{L}^T\ \underline{Y}_k^{i+N_d}\ \underline{Y}_x^{i+N_d} - \underline{L}^T\ \underline{Y}_h^{i+N_d}\}_{N-1},$$

where N indicates iteration number. From the above,

$$\underline{V}_{h,N}^i = [\underline{I} + \frac{1}{2m}\ \xi^i]^{-1}\ \underline{\mathcal{G}}^i,$$

$$\underline{Y}_{h,N}^i = \frac{1}{2}\ \underline{V}_{h,N}^i + \underline{\tilde{h}}(i),$$

$$\underline{\tilde{h}}(i-1) = \underline{V}_{h,N}^i + \underline{\tilde{h}}(i),$$

$$\underline{\mathcal{G}}^i = \underline{G}_{N-1}^i - \frac{1}{m}\ \xi^i\ \underline{\hat{h}}(i),\quad i = mt_f\ \ldots\ 3,2,1.$$

Starting values are taken to be unity satisfying initial conditions. After getting $\tilde{h}(t)$ and the PCBF values, (8.3c) is integrated forward as

$$\underline{V}^i_{x,N} = \frac{1}{m}\,[\underline{Z}^i\{\underline{V}^i_{x,N}\,\underline{E} + \underline{x}(i-1)\} + \underline{\tilde{R}}^i_{N-1}],$$

$$\underline{Z}^i = (\underline{A} - \underline{B}\,\underline{R}^{-1}\,\underline{B}^T\,\underline{Y}^i_k), \quad i = 1,2,3\ \ldots\ mt_f$$

$$\tilde{R}^i_{N-1} = [-\underline{B}\,\underline{R}^{-1}\,\underline{B}^T\,\underline{Y}^i_{h,N} + \underline{L}\,\underline{Y}^{i-N_d}_{x,N-1}].$$

The above can be solved as follows:

$$\underline{V}^i_{x,N} = \left[\underline{I} - \frac{\underline{Z}^i}{2m}\right]^{-1}\left[\frac{1}{m}\,\underline{Z}^i\,\underline{x}(i-1) + \frac{1}{m}\,\underline{\tilde{R}}^i_{N-1}\right],$$

$$\underline{Y}^i_{x,N} = \frac{1}{2}\,\underline{V}^i_{x,N} + \underline{x}(i-1),$$

$$\underline{x}(i) = \underline{V}^i_{x,N} + \underline{x}(i-1), \quad i = 1,2,3,\ \ldots\ mt_f$$

with $\underline{x}(0) = \underline{x}_\rho$.

If the state and adjoint equations are solved as above, $\underline{x}(t)$ converges to the optimal state $\overset{*}{\underline{x}}(t)$. $\overset{*}{u}(t)$, the optimal control may then be computed as follows:

$$\underline{Y}^i_{\overset{*}{u}} = -\underline{R}^{-1}\,\underline{B}^T\,[\underline{Y}^i_k\,\underline{Y}^i_x + \underline{Y}^i_h], \text{ or}$$

$$\overset{*}{\underline{u}}(i) = -\underline{R}^{-1}\,\underline{B}^T\,[\underline{K}(i)\,\underline{x}(i) + \underline{\tilde{h}}(i)].$$

8.4. Calculation of the performance index

In view of (8.1b)

$$\dot{J}(t) = \frac{1}{2}\,(\underline{x}^T\,\underline{Q}\,\underline{x} + \underline{u}^T\,\underline{R}\,\underline{u}).$$

Expanding this in single term PCBF with scale changed to $\hat{t} = mt$ as

$$\dot{J}(\hat{t}) = \underline{V}_j^i\,\theta_1(\hat{t});\quad J(t) = Y_j^i\,\theta_1(\hat{t});$$

$$\overset{*}{\underline{x}}(\hat{t}) = \underline{Y}_{\underset{x}{*}}^i\,\theta_1(\hat{t});\quad \overset{*}{u}(\hat{t}) = \underline{Y}_{\underset{u}{*}}^i\,\theta_1(\hat{t});$$

$$\underline{V}_j^i = \frac{1}{2m}\,[(\underline{Y}_{\underset{x}{*}}^i)^T\,\underline{Q}\,\underline{Y}_{\underset{x}{*}}^i + (\underline{Y}_{\underset{u}{*}}^i)^T\,\underline{R}\,\underline{Y}_{\underset{u}{*}}^i];$$

$$Y_j^i = \frac{1}{2}\,V_j^i + J(i-1);$$

$$J(i) = J(i-1) + V_j^i,\quad i = 1,2,\ \ldots\ m(t),$$

$$J(0) = 0.$$

8.5. Illustrative examples

Example 8.1.:

Let us consider the example of Aggarwal [C1,C2]

$$\dot{x}(t) = x(t) + x(t-1) + u(t),$$

$$x(t) = 1, \; -1 \leq t \leq 0,$$

$$J = \frac{1}{2} \int_0^2 (x^2 + u^2)\, dt .$$

Applying the Riccati transformation to the canonical equations the following scalar Riccati, adjoint and state equations are obtained:

$$\dot{K}(t) = K^2(t) + 2\,K(t) - 1, \; K(t) = 0, \; t \geq t_f \text{ since } G = 0;$$

$$\dot{\tilde{h}}_N(t) = (K(t) - 1)\, \tilde{h}_N(t) - K(t)\, x_{N-1}(t-1)$$

$$- K(t+1)\, x_{N+1}(t+1) - \tilde{h}_{N-1}(t+1), \; \tilde{h}(t) = 0, \; t \geq t_f ;$$

$$\dot{x}_N(t) = (1-K(t))\, x_N(t) - \tilde{h}_N(t) + x_{N-1}(t-1), \; x(0) = 1,$$

$$-1 \leq t \leq 0 .$$

The results are shown in Fig. 8.1 in comparison with those by other methods.

Example 8.2.:

Consider the system [C1,C2]

$$\dot{x}(t) = -x(t-1) + u(t) \; ;$$

$$x(t) = 1, \; -1 \leq t \leq 0 \; ;$$

$$J = \frac{1}{2} \int_0^1 (x^2 + u^2)\, dt .$$

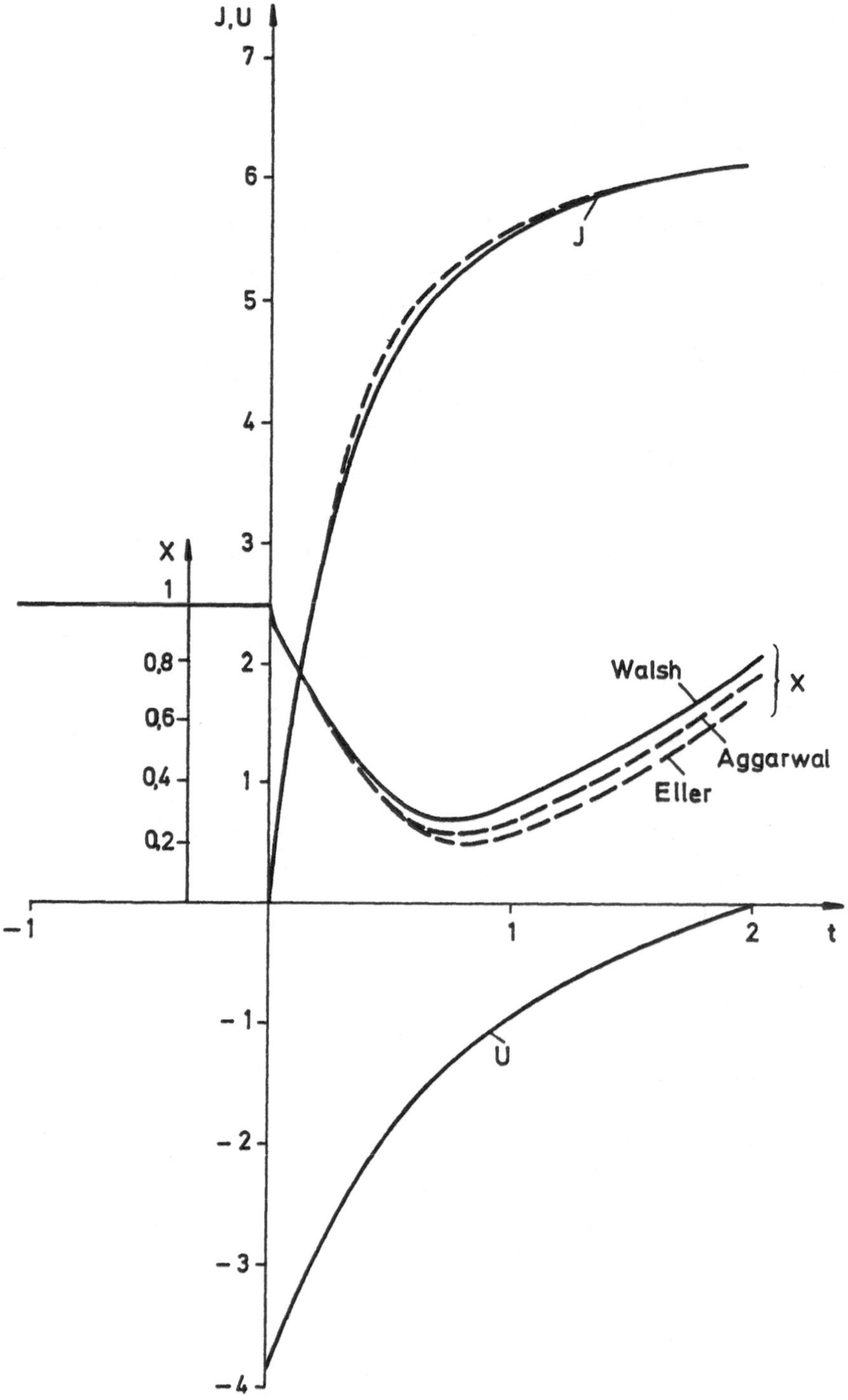

Fig. 8.1. Solution of Example 8.1

The scalar Riccati, adjoint and state equations in this case happen to be

$$\dot{K}(t) = K^2(t) - 1,\ K(t) = 0,\ t \geq t_f,$$

$$t_f = 1, \text{ since } G = 0;$$

$$\dot{\tilde{h}}_N(t) = K(t)\ \tilde{h}_N(t) + K(t)\ x_{N-1}(t-1) + K(t+1)\ x_{N-1}(t+1) + \tilde{h}_{N-1}(t+1)$$

$$\tilde{h}(t) = 0,\ t \geq t_f;$$

$$\dot{x}_N(t) = -K(t)\ x_N(t) - \tilde{h}_N(t) - x_{N-1}(t-1),\ x(0) = 1,\ -1 \leq t \leq 0.$$

The results computed from the above are shown in Fig. 8.2.

Example 8.3.:

Consider the system [C4-C10,C15]

$$\dot{x}(t) = x(t-1) + u(t),$$

$$x(t) = 1.0, -1 \leq t \leq 0,$$

with the performance index

$$J = \frac{1}{2} \int_0^2 (x^2 + u^2)\ dt.$$

The results obtained are given in Table 8.1. The results compare well with the solutions obtained by Banks et al. [C7].

The performance index

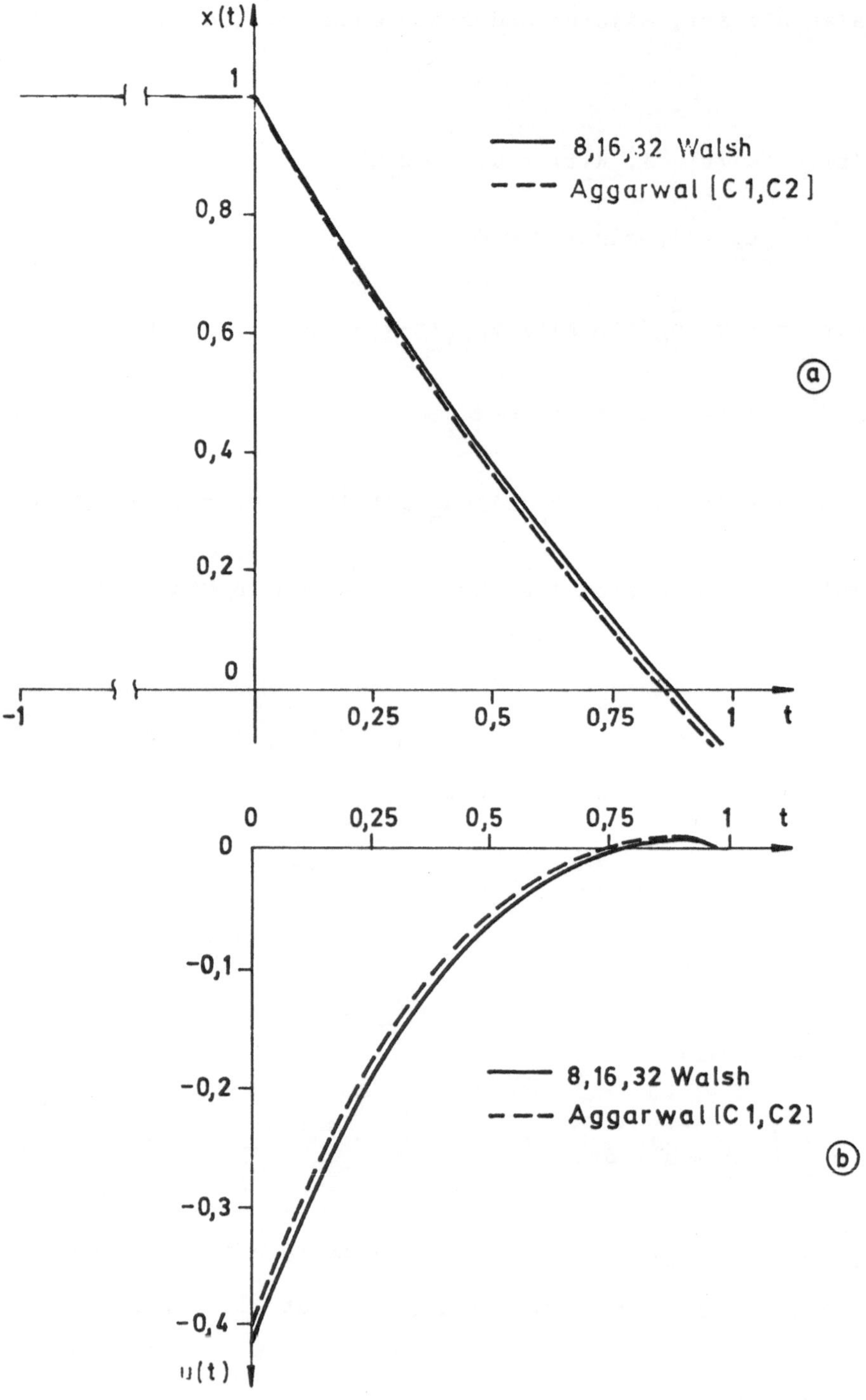

Fig. 8.2. Solution of Example 8.2

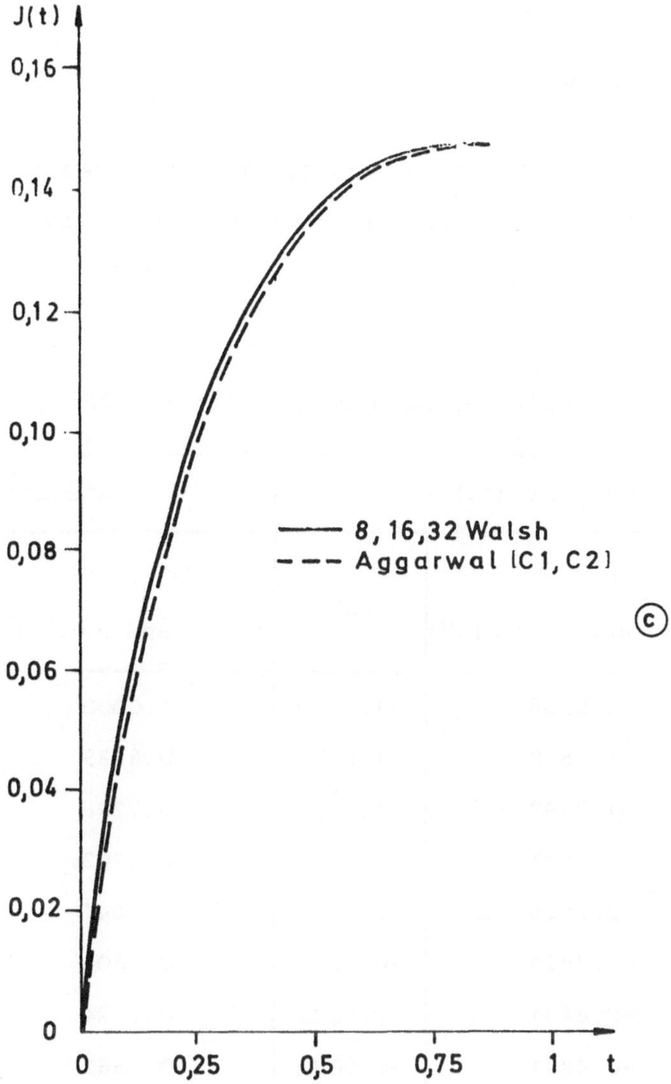

Fig. 8.2. Solution of Example 8.2 contd.

$J = 1.6497$ with $m = 100$,

$= 1.6504$ with $m = 10$.

Banks and Burns [C7] obtain J as 1.6419 with index $N_B = 20$. The solution by PCBF method converged within four iterations.

TABLE 8.1

Results of Example 8.3 with $m = 100$

t	Control u(t) by		State x(t) by	
	$N_B = 20$ Banks and Burns [C7]	PCBF	$N_B = 20$ Banks and Burns [C7]	PCBF
0.00	-1.9908	-1.9870	1.0000	1.0000
0.20	-1.6559	-1.6566	0.8339	0.8364
0.41	-1.3798	-1.3691	0.7296	0.7299
0.61	-1.1511	-1.1143	0.6763	0.6794
0.81	-0.9604	-0.9547	0.6641	0.6703
1.02	-0.7974	-0.7947	0.6804	0.6971
1.22	-0.6481	-0.6525	0.7124	0.7321
1.43	-0.4981	-0.5031	0.7562	0.7716
1.63	-0.3386	-0.3362	0.8174	0.8310
1.83	-0.1641	-0.1631	0.9057	0.9163
2.00	0.0	0.0	1.0091	1.0189

Example 8.4.:

Consider the example of Banks and Burns [C7]

$$\dot{x}(t) = x(t) + x(t-1) + u(t),$$

$$x(t) = 1{,}0, \; -1 \leq t \leq 0,$$

with performance index

$$J = 3/2 \; [x(2)]^2 + \frac{1}{2} \int_0^2 u(t)^2 \, dt.$$

The results obtained for the above problem are given in Table 8.2 The results compare well with the exact solution and solution by Banks et al. [C7].

The performance index J = 3.0876 with m = 100,

= 3.0889 with m = 10.

Banks and Burns [C7] obtain average J as 3.0833 using averaging approximation technique with the index $N_B = 20$. The PCBF solution converged within six iterations to the level of accuracy at which the integral squared error

$$\int_0^2 (x_N - x_{N-1})^2 \, dt \leq 4934 \times 10^{-8}.$$

TABLE 8.2

Results of Example 8.4 with m = 100

t	Control u(t)			state x(t) by
	Exact [C7]	$N_B = 20$ Banks et al. [C7]	PCBF	PCBF
0.00	-3.9787	-3.9734	-3.9730	1.0000
0.19	-3.1182	-3.1201	-3.1178	0.6772
0.39	-2.4083	-2.3973	-2.4079	0.4374
0.59	-1.8533	-1.8376	-1.8528	0.2845
0.80	-1.4014	-1.4082	-1.4008	0.2050
1.00	-1.0687	-1.0884	-1.0680	0.1988
1.20	-0.8750	-0.8604	-0.8747	0.2110
1.41	-0.7093	-0.6961	-0.7093	0.2004
1.61	-0.5807	-0.5678	-0.5810	0.1773
1.81	-0.4755	-0.4635	-0.4759	0.1506
2.00	-0.3932	-0.3842	-0.3937	0.1312

Example 8.5.:

Consider the example of Banks and Burns [C7]

$$\dot{x}(t) = x(t-1) + u(t),$$

$$x(t) = 1.0, \quad -1 \leq t \leq 0 .$$

With the performance index

$$J = 3/2 \, [x(3)]^2 + \frac{1}{2} \int_0^3 [u(t)]^2 \, dt.$$

The results obtained are given in Table 8.3 with the results obtained by Banks et al. [C7].

The value of performance index

$J = 1.7088$ with $m = 100$,

$= 1.7098$ with $m = 10$.

Banks and Burns obtain J as 1.7338 (average value).

TABLE 8.3

Results of Example 8.5 with $m = 100$

t	Control u(t) by			state x(t) by
	Banks and Burns [C7] $N_B = 20$	Exact [C7]	PCBF	PCBF
0.00	-1.9663	-1.9681	-1.9765	1.0000
0.30	-1.6518	-1.6656	-1.6630	0.7553
0.61	-1.3898	-1.3867	-1.3916	0.5933
0.91	-1.1731	-1.7775	-1.1799	0.5088
1.20	-1.0016	-1.0122	-1.0120	0.4633
1.50	-0.8309	-0.8435	-0.8403	0.4043
1.80	-0.6813	-0.6748	-0.6684	0.3530
2.10	-0.5761	-0.5623	-0.5538	0.3269
2.41	-0.5534	-0.5623	-0.5544	0.2950
2.72	-0.5529	-0.5623	-0.5550	0.2440
3.00	-0.5528	-0.5623	-0.5554	0.1851

Example 8.6.:

Consider the system [C2]

$$\dot{x}(t) = -ax(t-1) + u(t),$$

$$x(t) = 1, \; -1 \le t \le 0,$$

with

$$J = \frac{1}{2} \int_0^2 (x^2 + u^2)\, dt .$$

The related Riccati, adjoint and state equations are (with $a = 1$)

$$\dot{K}(t) = K^2(t) - 1, \; K(t) = 0, \; t \ge t_f ;$$

$$\dot{\tilde{h}}_N(t) = K(t)\, \tilde{h}_N(t) + K(t)\, x_{N-1}(t-1) + K(t+1)\, x_{N-1}(t+1)$$

$$+ \tilde{h}_{N-1}(t+1), \; \tilde{h}(t) = 0, \; t \ge t_f ;$$

$$\dot{x}_N(t) = -K(t)\, x_N(t) - \tilde{h}_N(t) - x_{N-1}(t-1),$$

$$x(0) = 1, \; x(t) = 1, \; -1 \le t \le 0.$$

Figures 8.3a-c show the optimal state, control and performance respectively.

Examples 8.7 and 8.8:

Example 8.6 with $a = 0.6$ and 0.8 with all other conditions remaining the same has been considered. Optimal controls computed

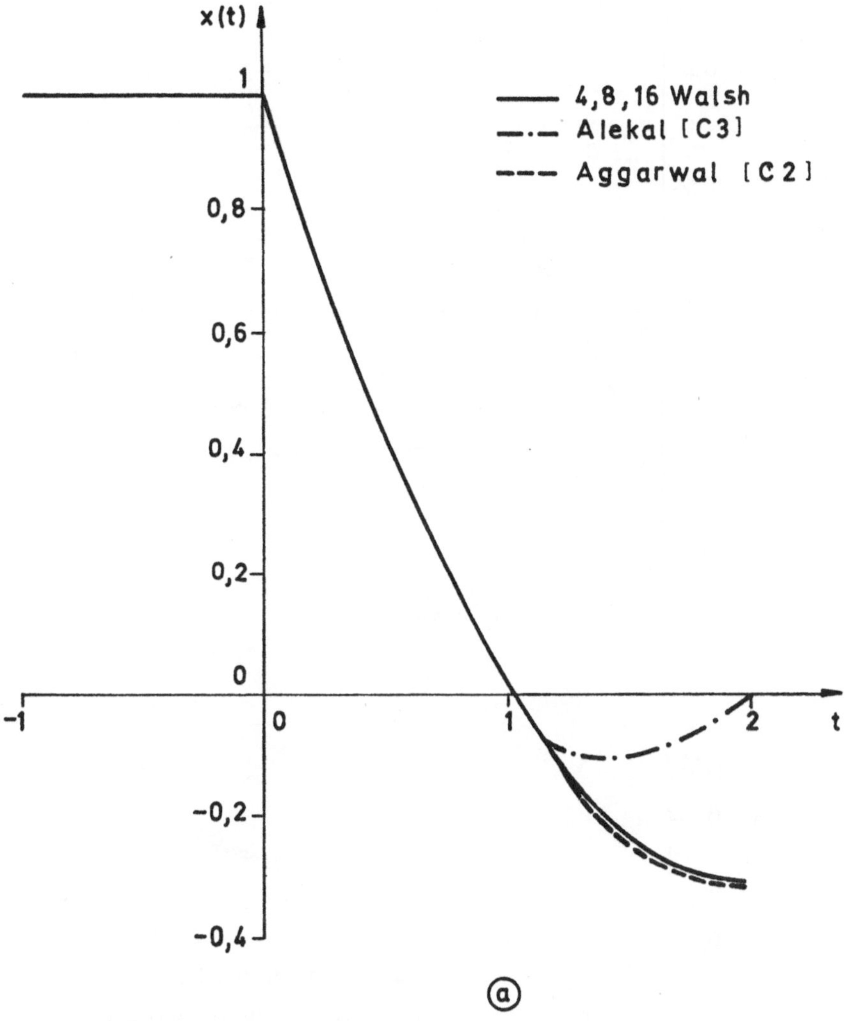

Fig. 8.3. Results of Example 8.6 (a = 1)

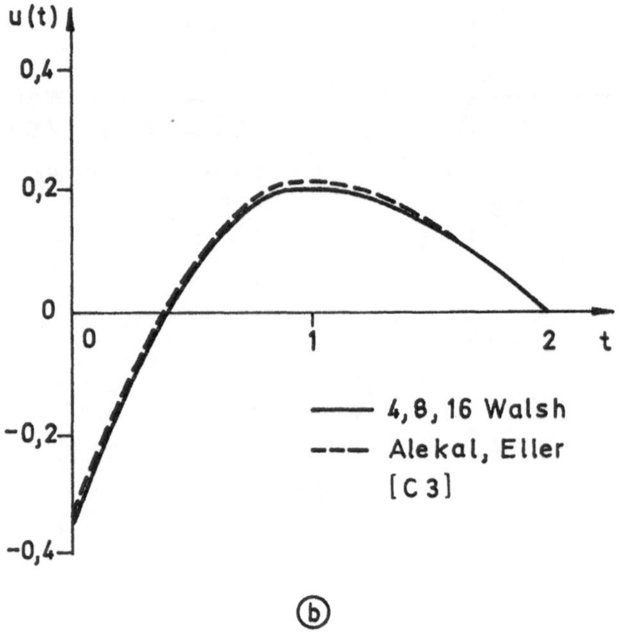

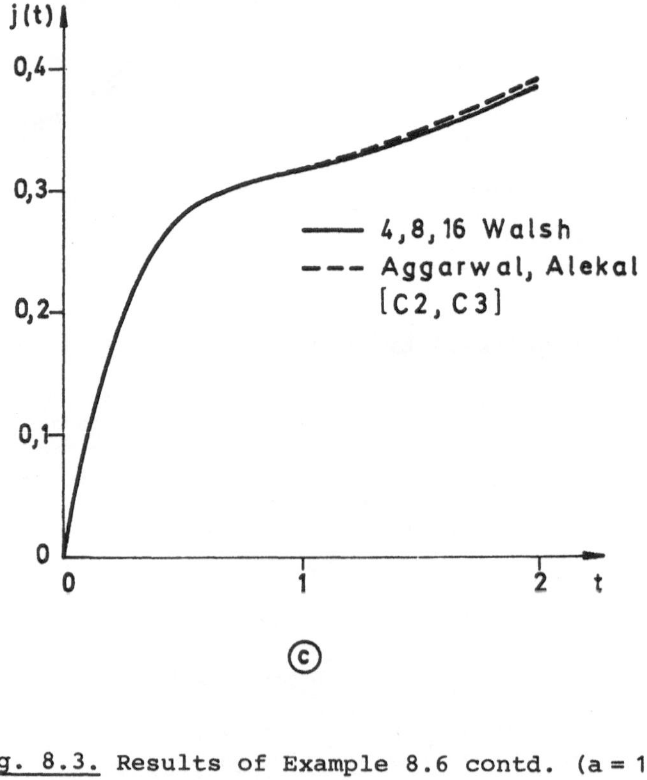

Fig. 8.3. Results of Example 8.6 contd. (a = 1)

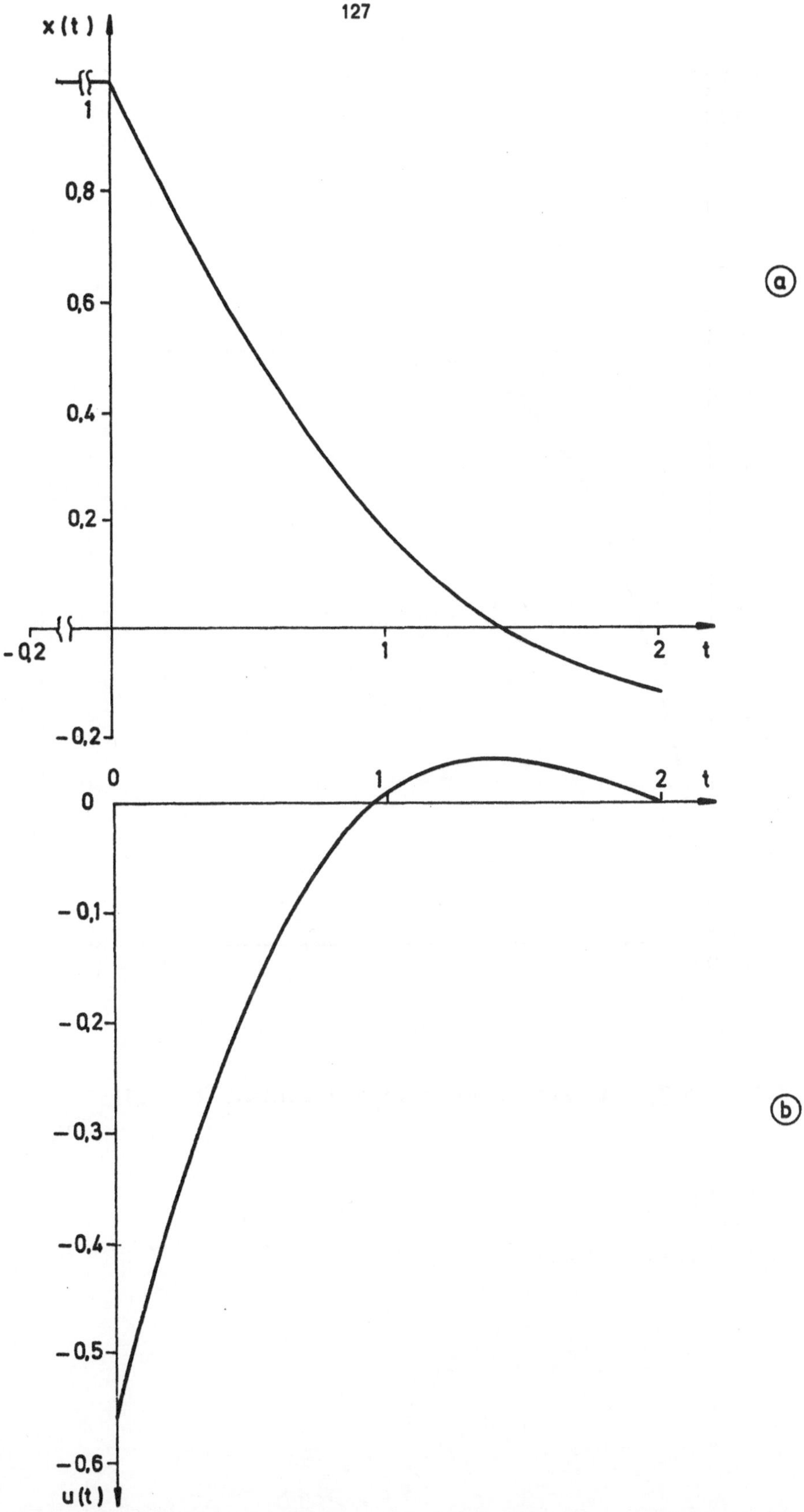

Fig. 8.4. Results of Example 8.7 (a = 0.6)

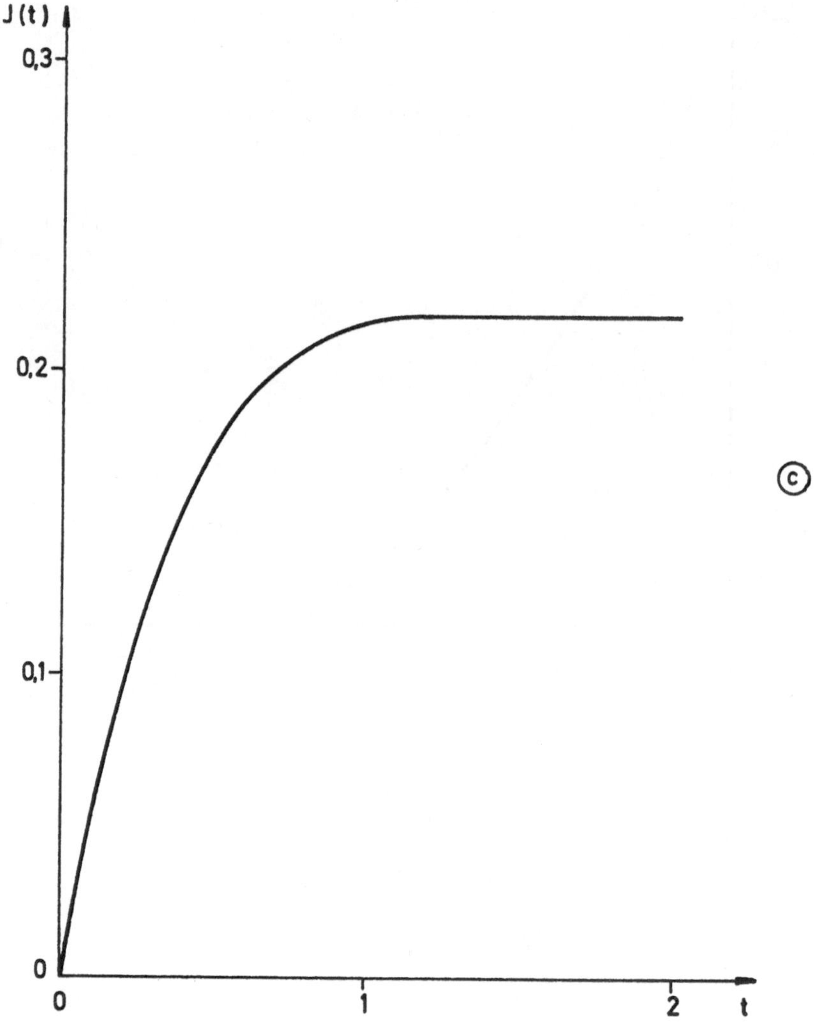

Fig. 8.4. Results of Example 8.7 contd. (a = 0.6)

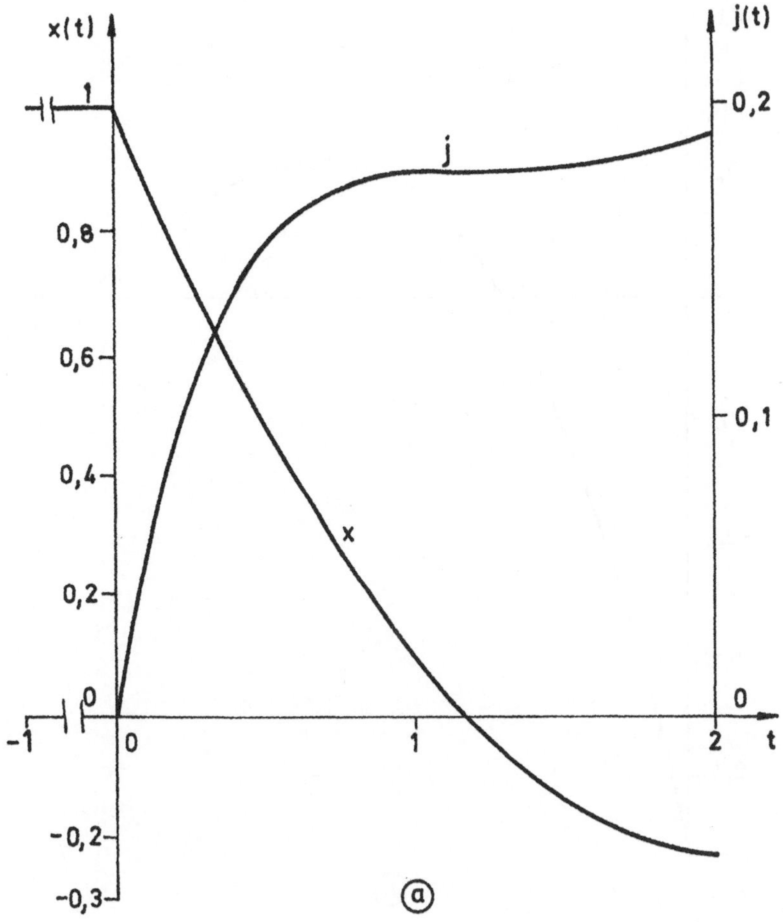

Fig. 8.5. Results of Example 8.8 (a = 0.8)

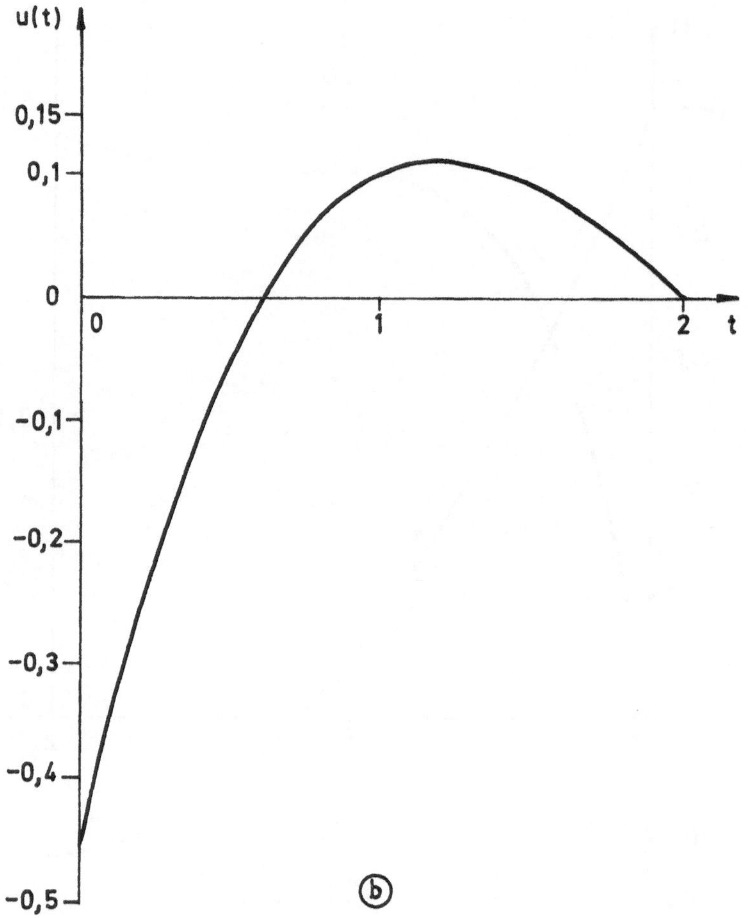

Fig. 8.5. Results of Example 8.8 contd. (a = 0.8)

in these cases by the PCBF method are shown in Figures 8.4 and 8.5 respectively.

8.6. The case of delays both in state and control

Consider a linear system with delays in both the state and control described by

$$\dot{\underline{x}}(t) = \underline{A}\underline{x}(t) + \underline{L}\underline{x}(t-\tau_s)+\underline{B}_1\underline{u}(t)+\underline{B}_2\underline{u}(t-\tau_c), \qquad (8.7a)$$

with initial data

$$\underline{x}(t_o) = \underline{x}_o, \qquad (8.7b)$$

$$\underline{x}(t) = \underline{x}_b(t), \quad t\varepsilon[t_o-\tau_s, t_o], \qquad (8.7c)$$

$$\underline{u}(t) = \underline{u}_b(t), \quad t\varepsilon[t_o-\tau_c, t_o]. \qquad (8.7d)$$

$\underline{x}$ is an n-vector of state; $\underline{u}$ is an r-vector of inputs; $\underline{A},\underline{L},\underline{B}_1$ and $\underline{B}_2$ are matrices of appropriate dimensions. τ_s and τ_c are delays in state and control respectively and the initial functions $\underline{x}_b(t)$ and $\underline{u}_b(t)$ are continuous in their respective intervals. The problem is to minimize

$$J = 1/2\ \underline{x}^T(t_f)\ \underline{G}\underline{x}(t_f) + 1/2 \int_o^{t_f} (\underline{x}^T\underline{Q}\underline{x} + \underline{u}^T\underline{R}\underline{u})\ dt,$$

where $\underline{Q}$ is a constant symmetric positive semi-definite matrix and $\underline{R}$ is a constant symmetric positive definite matrix. We wish to find a control $\underline{u}(t)$, $0 \leq 1 \leq t_f$, which, for the initial conditions as specified in (8.7b-d) for the fixed initial state $\underline{x}_o$, free final state $\underline{x}_f$ and the fixed terminal time t_f, minimizes the performance index J subject to (8.7a). The state and costate equations in the canonical form [C28] may be shown to be

$$\left.\begin{aligned}
\dot{\underline{x}}(t) &= \underline{\underline{A}}\underline{x}(t) + \underline{\underline{L}}\underline{x}(t-\tau_s) - \underline{\underline{B}}_1\underline{\underline{R}}^{-1}\underline{\underline{B}}_1^T\underline{p}(t) - \underline{\underline{B}}_1\underline{\underline{R}}^{-1}\underline{\underline{B}}_2^T\underline{p}(t+\tau_c) + \underline{\underline{B}}_2\underline{u}(t-\tau_c), \\
&\qquad t\in[t_o,\ t_o+\tau_c]; \qquad\qquad (a) \\
&= \underline{\underline{A}}\underline{x}(t) + \underline{\underline{L}}\underline{x}(t-\tau_s) - [\underline{\underline{B}}_1\underline{\underline{R}}^{-1}\underline{\underline{B}}_1^T + \underline{\underline{B}}_2\underline{\underline{R}}^{-1}\underline{\underline{B}}_2^T]\ \underline{p}(t) - \underline{\underline{B}}_1\underline{\underline{R}}^{-1}\underline{\underline{B}}_2^T\underline{p}(t+\tau_c) \\
&\qquad -\underline{\underline{B}}_2\underline{\underline{R}}^{-1}\underline{\underline{B}}_1^T\underline{p}(t-\tau_c),\ t\in[t_o+\tau_c,\ t_f-\tau_c]; \qquad (b) \\
&= \underline{\underline{A}}\underline{x}(t) + \underline{\underline{L}}\underline{x}(t-\tau_s) - [\underline{\underline{B}}_1\underline{\underline{R}}^{-1}\underline{\underline{B}}_1^T + \underline{\underline{B}}_2\underline{\underline{R}}^{-1}\underline{\underline{B}}_2^T]\underline{p}(t) - \underline{\underline{B}}_2\underline{\underline{R}}^{-1}\underline{\underline{B}}_1^T\ \underline{p}(t-\tau_c), \\
&\qquad t\in[t_f-\tau_c,\ t_f]; \qquad\qquad (c) \\
\dot{\underline{p}}(t) &= -\underline{\underline{A}}^T\underline{p}(t) - \underline{\underline{L}}^T\underline{p}(t+\tau_s) - \underline{\underline{Q}}\underline{x}(t),\ t\in[t_o,t_f]. \qquad (d)
\end{aligned}\right\}\quad (8.8)$$

Equations (8.8) satisfy the conditions in (8.7b-d). In addition $\underline{p}(t)=0$ for $t \geq t_f$. The optimal response $\underline{x}^*(t)$ and $\underline{p}^*(t)$ may be employed to get the optimal control.

$$\left.\begin{aligned}
\underline{u}^*(t) &= -\underline{\underline{R}}^{-1}\underline{\underline{B}}_1^T\underline{p}^*(t) - \underline{\underline{R}}^{-1}\underline{\underline{B}}_2^T\underline{p}^*(t+\tau_c), \\
&\qquad t\in[t_o,\ t_f-\tau_c]; \qquad (e) \\
&= -\underline{\underline{R}}^{-1}\underline{\underline{B}}_1^T\underline{p}^*(t),\ t\in[t_f-\tau_c,\ t_f]. \quad (f)
\end{aligned}\right\}\quad (8.8)$$

Equations (8.8) may be converted into initial value problems by an imbedding method. We also have

$$\underline{p}(t) = \underline{K}(t)\ \underline{x}(t) + \underline{\tilde{h}}(t).$$

Inserting the above into (8.8) we get

$$\dot{\underline{K}}(t) = \underline{K}(t)\underline{c}_1\underline{K}(t) - \underline{K}(t)\underline{A} - \underline{A}^T\underline{K}(t) - \underline{Q},\ t \in [t_o,\ t_o + \tau_c]$$

$$\text{and } \underline{K}(t) = \underline{G} \text{ for } t > t_f\ ; \qquad \text{(a)} \qquad (8.9)$$

$$= \underline{K}(t)\underline{c}_5\underline{K}(t) - \underline{K}(t)\underline{A} - \underline{A}^T\underline{K}(t) - \underline{Q},\ t \in [t_o + \tau_c,\ t_f]; \qquad \text{(b)}$$

$$\dot{\underline{\tilde{h}}}_N(t) = (\underline{K}(t)\underline{c}_1 - \underline{A}^T)\ \underline{\tilde{h}}_N(t) - \underline{K}(t)\{\underline{L}\underline{x}_{N-1}(t-\tau_s) + \underline{c}_3$$

$$[\underline{K}(t+\tau_c)\ \underline{x}_{N-1}(t+\tau_c) + \underline{\tilde{h}}_{N-1}(t+\tau_c)]$$

$$+ \underline{c}_4\ [\underline{K}(t-\tau_c)\ \underline{x}_{N-1}(t-\tau_c) + \underline{\tilde{h}}_{N-1}(t-\tau_c)]\}$$

$$-\underline{L}^T\underline{\tilde{h}}_{N-1}(t+\tau_s) - \underline{L}^T\underline{K}(t+\tau_s)\ \underline{x}_{N-1}(t+\tau_s),$$

$$t \in [t_o, t_o+\tau_c]; \qquad (8.9c)$$

$$= (\underline{K}(t)\underline{c}_5 - \underline{A}^T)\ \underline{\tilde{h}}_N(t) - \underline{K}(t)\{\underline{L}\underline{x}_{N-1}(t-\tau_s)$$

$$+ \underline{c}_3[\underline{K}(t+\tau_c)\ \underline{x}_{N-1}(t+\tau_c) + \underline{\tilde{h}}_{N-1}(t+\tau_c)]$$

$$+ \underline{c}_4[\underline{K}(t-\tau_c)\ \underline{x}_{N-1}(t-\tau_c) + \underline{\tilde{h}}_{N-1}(t-\tau_c)]\}$$

$$- \underline{L}^T\underline{K}(t+\tau_s)\ \underline{x}_{N-1}(t+\tau_s) - \underline{L}^T\underline{\tilde{h}}_{N-1}(t+\tau_s),$$

$$t \in [t_o + \tau_c,\ t_f - \tau_c]; \qquad (8.9d)$$

$$= (\underline{K}(t)\ \underline{C}_5 - \underline{A}^T)\ \tilde{\underline{h}}_N(t) - \underline{K}(t)\{\underline{L}\underline{x}_{N-1}(t-\tau_s)$$

$$+ \underline{C}_4[\underline{K}(t-\tau_c)\ \underline{x}_{N-1}(t-\tau_c) + \tilde{\underline{h}}_{N-1}(t-\tau_c)]\}$$

$$- \underline{L}^T\underline{K}(t+\tau_s)\ \underline{x}_{N-1}(t+\tau_s) - \underline{L}^T\tilde{\underline{h}}_{N-1}(t+\tau_s),$$

$$\tilde{\underline{h}}(t) = 0 \text{ for } t \geq t_f \text{ and } t \in [t-\tau_c, t_f]; \qquad (8.9e)$$

$$\dot{x}_N(t) = (\underline{A}-\underline{C}_1\underline{K}(t))\,\underline{x}_N(t) + \underline{L}\underline{x}_{N-1}(t-\tau_s) - \underline{C}_1\tilde{\underline{h}}_N(t)$$

$$- \underline{C}_3[\underline{K}(t+\tau_c)\ \underline{x}_{N-1}(t+\tau_c) + \tilde{\underline{h}}_N(t+\tau_c)]$$

$$+\underline{B}_2\underline{u}(t-\tau_c), \quad t \in [t_o, t_o+\tau_c]; \qquad (8.9f)$$

$$= (\underline{A}-\underline{C}_5\underline{K}(t))\,\underline{x}_N(t) + \underline{L}\underline{x}_{N-1}(t-\tau_s) - \underline{C}_5\ \tilde{\underline{h}}_N(t)$$

$$- \underline{C}_3[\underline{K}(t+\tau_c)\ \underline{x}_{N-1}(t+\tau_c) + \tilde{\underline{h}}_N(t+\tau_c)]$$

$$- \underline{C}_4[\tilde{\underline{h}}_N(t-\tau_c) + \underline{K}(t-\tau_c)\ \underline{x}_{N-1}(t-\tau_c)],$$

$$t \in [t_o+\tau_c, t_f-\tau_c]; \qquad (8.9g)$$

$$= (\underline{A}-\underline{C}_5\underline{K}(t))\ \underline{x}_N(t) + \underline{L}\underline{x}_{N-1}(t-\tau_s) - \underline{C}_5\tilde{\underline{h}}_N(t)$$

$$- \underline{C}_4[\underline{K}(t-\tau_c)\ \underline{x}_{N-1}(t-\tau_c) + \tilde{\underline{h}}_N(t-\tau_c)],$$

$$t \in [t_f-\tau_c, t_f], \qquad (8.9h)$$

where $\underline{C}_1 = \underline{B}_1\underline{R}^{-1}\underline{B}_1^T$, $\underline{C}_2 = \underline{B}_2\underline{R}^{-1}\underline{B}_2^T$, $C_3 = \underline{B}_1\underline{R}^{-1}\underline{B}_2^T$,

$\underline{C}_4 = \underline{B}_2\underline{R}^{-1}\underline{B}_1^T$ and $\underline{C}_5 = \underline{C}_1 + \underline{C}_2$.

The Riccati equations (8.9a) and (8.9b) are solved first. Then, equations (8.9c-e) are solved by backward integration to obtain the adjoint vector $\underline{\tilde{h}}(t)$ and equations (8.9f-h) are integrated forward in time to obtain $\underline{x}(t)$. These equations are solved iteratively and subscript N denotes the iteration number.

8.7. Solution of Riccati differential equation

To solve equation (8.9), equations (8.8) are first written in state-space form as:

$$\begin{bmatrix} \dot{\underline{x}}(t) \\ \dot{\underline{p}}(t) \end{bmatrix} = \begin{bmatrix} \underline{A} & -\underline{C}_5 \\ -\underline{Q} & -\underline{A}^T \end{bmatrix} \begin{bmatrix} \underline{x}(t) \\ \underline{p}(t) \end{bmatrix} + \text{delay terms.} \tag{8.10}$$

Let $\eta(t_f,t)$ be the transition matrix of equation (3.10) (omitting delay terms) and $\eta(t_f,t_f) = \underline{I}$.

$\eta(t_f,t)$ may be decomposed into the following form:

$$\underline{\eta}(t_f,t) = \begin{bmatrix} \underline{\eta}_{11}(t_f,t) & \underline{\eta}_{12}(t_f,t) \\ \underline{\eta}_{21}(t_f,t) & \underline{\eta}_{22}(t_f,t) \end{bmatrix}. \tag{3.11}$$

We may thus write

$$\begin{bmatrix} \underline{x}(t_f) \\ \underline{p}(t_f) \end{bmatrix} = \begin{bmatrix} \underline{\eta}_{11}(t_f,t) & \underline{\eta}_{12}(t_f,t) \\ \underline{\eta}_{21}(t_f,t) & \underline{\eta}_{22}(t_f,t) \end{bmatrix} \begin{bmatrix} \underline{x}(t) \\ \underline{p}(t) \end{bmatrix}, \tag{8.12}$$

giving

$$\underline{K}(t) = \underline{\Omega}_{22}^{-1}(t_f,t)\ \underline{\Omega}_{21}(t_f,t),$$

$$t \in [t_o + t_c,\ t_f]. \tag{8.13}$$

Similarly $\underline{K}(t)$ in (8.9a) may be obtained from the transition matrix of

$$\begin{bmatrix} \dot{\underline{x}}(t) \\ \dot{\underline{p}}(t) \end{bmatrix} = \begin{bmatrix} \underline{A} & -\underline{C}_1 \\ -\underline{Q} & -\underline{A}^T \end{bmatrix} \begin{bmatrix} \underline{x}(t) \\ \underline{p}(t) \end{bmatrix} + \text{delay terms}, \tag{8.14}$$

which is obtained from (8.8a) and (8.8d) in the interval $t \in [t_o, t_o+\tau_c]$, on the lines of (8.11-8.13).

Equations (8.9c-h) containing delayed and advanced arguments should now be solved in the respective intervals with the PCBF approach.

8.8. Solution of adjoint and state equations

In the backward integration of the Riccati equation, the procedure described in equations (8.10-8.14) will now take the algebraic form after a change of time scale towards normalization of the time segment $\frac{1}{m}$ for a positive integer m

$$\begin{bmatrix} \dot{\underline{x}} \\ \dot{\underline{p}} \end{bmatrix} = -\frac{1}{m} \begin{bmatrix} \underline{A} & -\underline{C}_5 \\ -\underline{Q} & -\underline{A}^T \end{bmatrix} \begin{bmatrix} \underline{x} \\ \underline{p} \end{bmatrix}, \quad \begin{matrix} \underline{x}(0) = \underline{I}, \\ \underline{p}(0) = 0.0, \text{ and} \end{matrix} \tag{8.15}$$

the recursive equations take the form:

$$\underline{V}_{xp}^{(j)} = \left[\underline{I} + \frac{\underline{H}_b}{2m}\right]^{-1} \frac{\underline{H}_b}{m} \underline{x}_{xp}(j-1),$$

$$\underline{Y}_{xp}^{(j)} = \frac{1}{2} \underline{V}_{xp}^{(j)} + \underline{x}_{xp}(j-1), \qquad (8.16)$$

$$\underline{x}_{xp}(j) = \underline{V}_{xp}^{(j)} + \underline{x}_{xp}(j-1),$$

$$j = m.t_f \; \ldots \; m.(t_o + \tau_c),$$

where

$$\begin{bmatrix} \dot{\underline{x}} \\ \dot{\underline{p}} \end{bmatrix} = \underline{V}_{xp} \; \theta_1(\hat{t}); \qquad \begin{bmatrix} \underline{x} \\ \underline{p} \end{bmatrix} = \underline{Y}_{xp} \; \theta_1(\hat{t});$$

$$\underline{x}_{xp} = \begin{bmatrix} \underline{x} \\ \underline{p} \end{bmatrix} = \eta(t_f, t),$$

and

$$\underline{H}_b = \begin{bmatrix} \underline{A} & -\underline{c}_5 \\ -\underline{Q} & -\underline{A}^T \end{bmatrix}.$$

The solution is in the interval $[t_o + \tau_c, \; t_f]$ with the condition $K(t_f) = 0, t \geq t_f$. Next we consider (8.9) with the results obtained from (8.16) in the interval $[t_o, \; t_o + \tau_o]$. $\underline{\eta}(t_f, \; t)$ can be found by taking

$$\underline{H}_a = \begin{bmatrix} \underline{A} & -\underline{C}_1 \\ -\underline{Q} & -\underline{A}^T \end{bmatrix}, \tag{8.17}$$

and $\underline{K}(t_o + \tau_c)$ from the solution of equation (8.16).

The solution of Riccati equations (8.9) can be calculated from $\underline{\eta}(t_f, t)$ by using (8.13). The solution of Riccati equation is used in (8.9c-h) to solve them iteratively. At first the adjoint equations (8.9c-e) are solved by backward integration with $\underline{\tilde{h}}(t_f) = 0$ for $t \geq t_f$. We express $\underline{K}(\hat{t})$, $\underline{K}(\hat{t}\pm\bar{\theta})$, $\underline{x}(\hat{t}\pm\bar{\theta})$, $\dot{\underline{\tilde{h}}}(\hat{t})$, $\underline{\tilde{h}}(\hat{t})$, $\underline{x}_b(\hat{t})$, and $\tilde{h}(\hat{t}\pm\bar{\theta})$ in single term PCBF in the ith interval with $\bar{\theta}$ as any known delay or advance as:

$$\underline{K}(\hat{t}) = \underline{K}_i\ \theta_1(\hat{t})\ \underline{x};\ \underline{K}(\hat{t}\pm\bar{\theta}) = \underline{K}_{i\pm\alpha}\ \theta_1(\hat{t});$$

$$\underline{x}(\hat{t}) = \underline{Y}_i\ \theta_1(\hat{t});\ \underline{x}(\hat{t}\pm\bar{\theta}) = \underline{Y}_{i\pm\alpha}\ \theta_1(\hat{t});$$

$$\dot{\underline{x}}(\hat{t}) = \underline{V}_i\ \theta_1(\hat{t});\ \underline{\hat{h}}(\hat{t}) = \underline{Y}h_i\ \theta_1(\hat{t});$$

$$\underline{\tilde{h}}(\hat{t}\pm\bar{\theta}) = \underline{Y}h_{i\pm\alpha}\ \theta_1(\hat{t});\ \dot{\underline{\hat{h}}}(\hat{t}) = \underline{V}h_i\ \theta_1(\hat{t});$$

$$\underline{x}_b(t) = -\underline{Y}_i\ \theta_1(t),\ i = -m(t_o - \tau_s)\ \ldots\ m.t_o. \tag{8.18}$$

and $\alpha = m.\bar{\theta}$.

Equation (8.9e) can be written in each of the 1/m long normalised intervals for backward integration in PCBF (single term approximation) as:

$$\underline{V}\underline{h}_{i,N}\ \theta_1(\hat{t}) = -\frac{1}{m}\{\underline{\xi}_i[\underline{V}\underline{h}_{i,N}\ \underline{E} + \underline{\tilde{h}}(i+1)] + G_{i,N-1}\}\ \theta_1(\hat{t}). \tag{8.19}$$

where

$\underline{\xi}_i = (\underline{K}_i \underline{c}_5 - \underline{A}^T)$, $\underline{\tilde{h}}(i)$ = discrete time value of $\underline{\tilde{h}}(t)$, and

$$\underline{G}_{i,N-1} = \{\underline{K}_i(\underline{LY}_{i-\alpha_1} + \underline{c}_4 \underline{K}_{i-\alpha_2} \underline{Y}_{i-\alpha_2} + \underline{Yh}_{i-\alpha_2}) - \underline{L}^T \underline{K}_{i+\alpha_1} \underline{Y}_{i+\alpha_1}$$

$$- \underline{L}^T \underline{Yh}_{i+\alpha_1}\} N-1,$$

N being the iteration number $\alpha_1 = m.\tau_s$, $\alpha_2 = m.\tau_c$, $K_i = 0$ $i > m.t_f$ and $Yh_i = 0$ $i > m.t_f$.

Solving (8.19) we get

$$\left.\begin{aligned} \underline{Vh}_{i,N} &= [1 + \frac{1}{2m} \underline{\xi}_i]^{-1} G_i, \\ \underline{Yh}_{i,N} &= \frac{1}{2} \underline{Vh}_{i,N} + \underline{\tilde{h}}(i+1), \\ \underline{\tilde{h}}(i)_N &= \underline{Vh}_{i,N} + \underline{\tilde{h}}(i+1), \end{aligned}\right\} \quad (8.20)$$

where

$$\underline{G}_i = \underline{G}_{i,N-1} - \frac{1}{m} \underline{\xi}_i \underline{\tilde{h}}(i+1),$$

$$i = m.t_f \ldots m.(t_o - \tau_c) \text{ and}$$

$$\underline{\tilde{h}}(mt_f + 1) = 0.$$

In a similar fashion equations (8.8c) and (8.8d) are solved in their respective intervals. Initial starting values can be assumed as unity satisfying the given initial conditions.

After getting the adjoint vectors, equations (8.9f-h) may be solved for the state. Equation (8.9f) can be written in each of the $\frac{1}{m}$ long normalised intervals for forward integration as: (i.e. $\hat{t} = m.t.$)

$$\underline{V}_{i,N} = \frac{1}{m}\,[\underline{Z}_i\{\underline{V}_{i,N}\,\underline{E} + \underline{x}(i-1)\} + \tilde{R}_{i,N-1}],$$

where

$$\underline{Z}_i = (\underline{A} - \underline{C}_5\underline{K}_i),\ i = 1,2,3\ \ldots\ m.\tau_s,$$

$$\text{and}\quad \tilde{R}_{i,N-1} = [\underline{LY}_{i-\alpha_1,N-1} - \underline{C}_5\ \underline{Yh}_{i,N} - \underline{C}_4] \tag{8.21}$$

$$\{\underline{K}_{i-\alpha_2},\ \underline{Y}_{i-\alpha_2,N-1} + \underline{Yh}_{i-\alpha_2,N}\}.$$

Solving (8.21) we get

$$\left.\begin{aligned} \underline{Y}_{i,N} &= \frac{1}{2}\,\underline{V}_{i,N} + \underline{x}(i-1), \\ \underline{x}(i) &= \underline{V}_{i,N} + \underline{x}(i-1), \end{aligned}\right\} \tag{8.22}$$

$$i = 1,2\ \ldots\ m.\tau_s\ \text{and}\ \underline{x}(0) = \underline{x}_0.$$

Similarly equations (8.8b) and (8.8c) can be solved for other intervals.

Equations (8.9c-h) may be solved iteratively and finally the solution $\underline{x}(t)$ converges to $\underline{x}^*(t)$ giving the optimal control $\underline{u}^*(t)$. The optimal control is calculated as follows:

$$\bar{\underline{u}}_i = -\underline{R}^{-1} [\underline{B}_1^T\{(\underline{K}_i\underline{Y}_i + \underline{Y}\underline{h}_i\} + \underline{B}_2^T \{K_{i+\alpha_2}. \underline{Y}_{i+\alpha_2} + \underline{Y}\underline{h}_{i+\alpha_2}\},$$

$$t \in [t_o, t_f - \tau_c], \; i = m.t_o \ldots m.(t_f - \tau_c),$$

$$= -\underline{R}^{-1} \underline{B}_1^T[\underline{K}_i\underline{Y}_i + \underline{Y}\underline{h}_i]. \quad t \in [t_f - \tau_c, t_f] \text{ and}$$

$$i = m.(t_{f.} - \tau_c) \ldots m.t_f.$$

$$\underline{u}^*(t_i) = -\underline{R}^{-1} [\underline{B}_1^T\{\underline{K}(t_i) \; \underline{x}(t_i) + \tilde{\underline{h}}(t_i)\} + \underline{B}_2^T\{\underline{K}(t_i+\tau_c) \; \underline{x}(t_i+\tau_c)$$

$$+ \tilde{\underline{h}}(t_i+\tau_c)\}], \quad t \in [t_o, t_f - \tau_c] \quad i = m.t_o \ldots m(t_f - \tau_c)$$

$$= -\underline{R}^{-1} \underline{B}_1^T[\underline{K}(t_i) \; \underline{x}(t_i) + \tilde{\underline{h}}(t_i)], \quad t \in [t_f - \tau_c, t_f],$$

$$i = m.(t_f - \tau_c) \ldots m.t_f \tag{8.23}$$

and $\underline{u}^*(t_i)$ and $\bar{\underline{u}}_i$ represent the optimal discrete time value and PCBF (block pulse) value of the control function.

8.9. Calculation of performance index

Differentiating equation (8.1b),we get

$$\dot{J}(t) = \frac{1}{2} (\underline{x}^T\underline{Q}\underline{x} + \underline{u}^T\underline{R}\underline{u}). \tag{8.24}$$

Now equation (8.20) is represented by single term PCBF series as

$$(V_j)_i = \frac{1}{2m}\,[\underline{Y}_i^T\,\underline{Q}\underline{Y}_i + \underline{\bar{u}}_i\,\underline{R}\,\underline{\bar{u}}_i^T]. \tag{8.25}$$

Solving (8.21)

$$\left.\begin{aligned} (Y_j)_i &= \frac{1}{2}\,(V_j)_i + J(i-1), \\ J(i) &= (V_j)_i + J(i-1), \end{aligned}\right\} \tag{8.26}$$

$$i = 1,\ 2\ \ldots\ m.t_f.$$

where

$$J(\hat{t}) = Y_{j,i}\ \theta_1(\hat{t});\ \dot{J}(\hat{t}) = V_{j,i}\ \theta_1(\hat{t});$$

$$\underline{x}(\hat{t}) = \underline{Y}_i\ \theta_1(\hat{t});$$

$$\underline{u}(\hat{t}) = \underline{\bar{u}}_i\ \theta_1(\hat{t});\ J(\hat{t}) = J(i)\ \text{and}$$

$$J(0) = 0\,.$$

8.10. Numerical examples

EXAMPLE 8.9:

For a system described by [C11,C46]

$$\left.\begin{aligned} \dot{x}(t) &= -x(t) + u(t) - B_2u(t-\tau_c), \\ x(0) &= 1.0, \\ u(t) &= 0, \quad t \in [-\tau_c, 0], \end{aligned}\right\} \tag{8.27}$$

minimise

$$J = \int_0^1 (x^2(t) + u^2(t))\, dt.$$

Equation (8.27) is considered with following values of B_2 and τ_c

$B_2 = 0$

$= \sqrt{0.5}$ $\quad \tau_c = 2/3$

$= 0.5$ $\quad = 5/12$

$= 0.5$ $\quad = 2/3$

Comparing (8.27) with equations (8.1) we get

$A = -1.0,\ L = 0.0,\ B_1 = 1.0,\ R = 0.5,\ Q = 1.0,\ t_o = 0,$

$G = 0,\ t_f = 1.$

With $B_2 = 0.5$ the values of c_1, c_2, c_3, c_4, and c_5 become 2.0, 0.5, -1.0, -1.0, and 2.5 respectively.

We get the following scalar Riccati, adjoint and state equations: k(t) become a scalar k(t)

$$\dot{k}(t) = 2.5\, k^2(t) + 2k(t) - 1.0,$$

$$k(t) = 0,\ t \geq t_f \quad t \in [t_o + \tau_c,\ t_f], \tag{8.28a}$$

$$\dot{k}(t) = 2k^2(t) + 2k(t) - 1.0, \quad t \in [t_o,\ t_o + \tau_c] \tag{8.28b}$$

$$\left.\begin{aligned}
\dot{\tilde{h}}_N(t) &= (2k(t)+1.0)\ \tilde{h}_N(t) - k(t)\{-k(t+\tau_c) \\
&\quad x_{N-1}(t+\tau_c) - \tilde{h}_{N-1}(t+\tau_c) - k(t-\tau_c) \\
&\quad x_{N-1}(t-\tau_c) - \tilde{h}_{N-1}(t-\tau_c)\},\ t\,\varepsilon\,[t_o,\ t_o+\tau_c] \\
&= (2.5\ k(t)+1.0)\ \tilde{h}_N(t) - k(t)\ \{-[k(t)+\tau_c) \\
&\quad x_{N-1}(t+\tau_c) + \tilde{h}_{N-1}(t+\tau_c)] - [k(t-\tau_c) \\
&\quad x_{N-1}(t-\tau_c) + \tilde{h}_{N-1}(t-\tau_c)]\},\quad t\,\varepsilon\,[t_o+\tau_c,\ t_f-\tau_c] \\
&= (2.5\ k(t)+1.0)\ \tilde{h}_{N-1}(t) - k(t)\{-k(t-\tau_c) \\
&\quad x_{N-1}(t-\tau_c) + \tilde{h}_{N-1}(t-\tau_c)]\},\ t\,\varepsilon\,[t_f-\tau_c,\ t_f] \\
\tilde{h}(t) &= 0,\ \text{for}\ t \geq t_f.
\end{aligned}\right\}\quad (8.29)$$

$$\left.\begin{aligned}
\dot{x}_N(t) &= (-1.0-2.0\ k(t))\ x_N(t) - 2.0\ \tilde{h}_N(t) + [k(t+\tau_c) \\
&\quad x_{N-1}(t+\tau_c) + \tilde{h}_N(t+\tau_c)],\ t\,\varepsilon\,[t_o,\ t_o+\tau_c], \\
&= (-1.0-2.5\ k(t))\ x_N(t) - 2.5\ \tilde{h}_N(t) + [k(t+\tau_c) \\
&\quad x_{N-1}(t+\tau_c) + \tilde{h}_N(t+\tau_c) + k(t-\tau_c) \\
&\quad x_{N-1}(t-\tau_c) + \tilde{h}_N(t-\tau_c)],\ t\,\varepsilon[t_o+\tau_c,\ t_f-\tau_c] \\
&= (-1.0-2.5\ k(t))\ x_N(t) - 2.5\ \tilde{h}_N(t) + k(t-\tau_c) \\
&\quad x(t-\tau_c) + \tilde{h}(t-\tau_c).\ t\,\varepsilon\,[t_f-\tau_c,\ t_f].
\end{aligned}\right\}\quad (8.30)$$

Equations (8.28) may be solved directly by single term PCBF approach. Normalising (8.28) with $\hat{t} = mt$ and changing the sign of left hand side for backward integration.

$$\left.\begin{aligned}\dot{k}(\hat{t}) &= \frac{-2.5}{m} k^2(\hat{t}) - \frac{2}{m} k(\hat{t}) + \frac{1.0}{m},\\ \dot{k}(\hat{t}) &= \frac{-2.0}{m} k^2(\hat{t}) - \frac{2}{m} k(\hat{t}) + \frac{1.0}{m},\end{aligned}\right\} \qquad (8.31)$$

with the expansions in PCBF, viz:

$$\dot{k}(\hat{t}) = V_k\ \theta_1(\hat{t}) \text{ and } k(\hat{t}) = Y_k\ \theta_1(\hat{t}).$$

We get the following quadratic equations for (8.31)

$$\frac{2.5}{4}\ [(V_k)_i]^2 + (V_k)_i\ [m + 2.5\ k(i) + 1.0]$$

$$+ [2.5\ k^2(i) + 2k(i) - 1.0] = 0, \qquad (8.32a)$$

$$i = m.t_f\ \ldots\ m(t_o + \tau_c),$$

$$\frac{2.0}{4}\ [(V_k)_i]^2 + (V_k)_i\ [m + 2.0\ k(i) + 1.0]$$

$$+ [2.5\ k^2(i) + 2k(i) - 1.0] = 0 \qquad (8.32b)$$

$$i = m(t_o + \tau_c)\ \ldots\ m(t_o + \tau_c).$$

$k(i)$ represents discrete time values of $k(t)$ and $k(m.t_f) = 0.0$ since $G = 0$.

At each step

$$(Y_k)_i = \frac{1}{2}\ (Y_k)_i + k(i),$$

$$k(i) = (V_k)_i + k(i).$$

Then equations (8.29) and (8.30) may be solved iteratively by single term PCBF approach. Equations (8.23) and (8.26) may be used to get the control and the performance index. Figures 8.6 and 8.7 show the control, state and performance index calculated with $m = 10$ and $m = 100$. They compare well with the results of Budėlis et al. [C11].

EXAMPLE 8.10:

Consider the example of Soliman and Ray [C45,C46]

$$\dot{x}(t) = -x(t) + x(t-1/3) + u(t) - 0.5\, u(t-2/3), \tag{8.33}$$

$$x(t) = 1.0, \quad t \in [-1/3, 0],$$

$$u(t) = 0, \quad t \in [-2/3, 0]$$

with the performance index

$$J = \frac{1}{2} \int_0^1 (x^2(t) + u^2(t)\; dt .$$

The results obtained are shown in Fig. 8.8 ($m = 100$).

8.11. Minimum energy control of time delay systems [W36]

Minimum energy problems with zero terminal constraint are of considerable importance. Sawaragi et al. [C42,C43] suggest approximate sensitivity methods for systems with small delay in which the delay system is treated as a non-delay system with enlarged dimension.

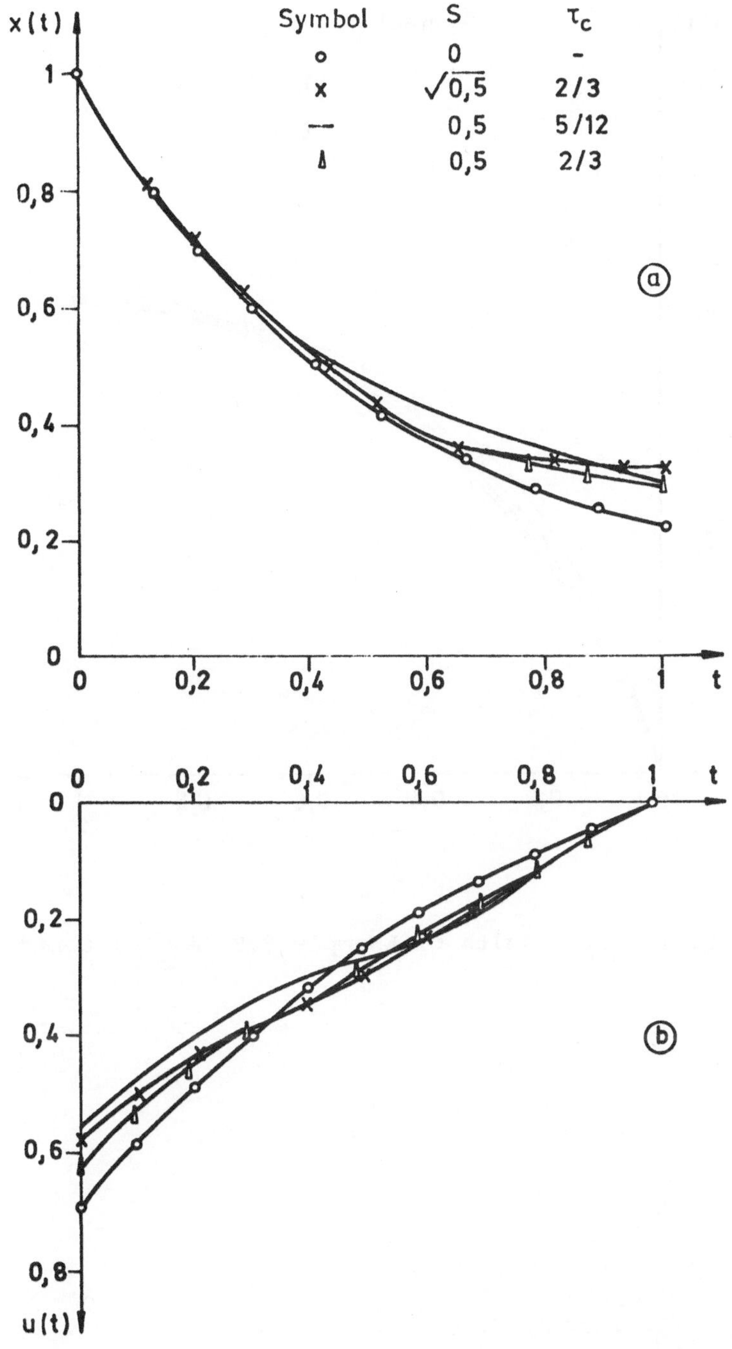

Fig. 8.6. Results of Example 8.9 (m = 10)

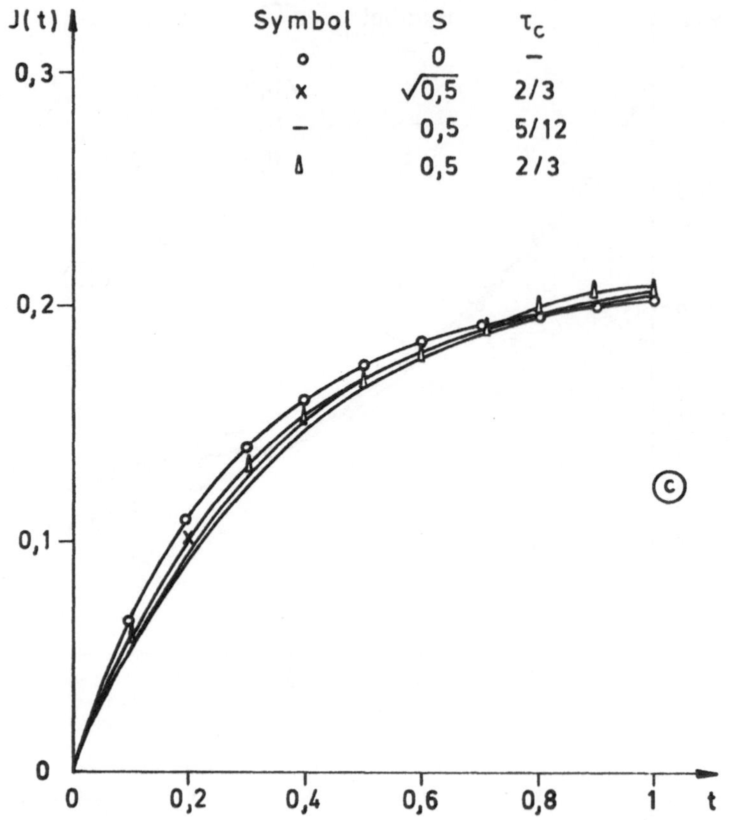

Fig. 8.6. Results of Example 8.9 (m = 10) contd.

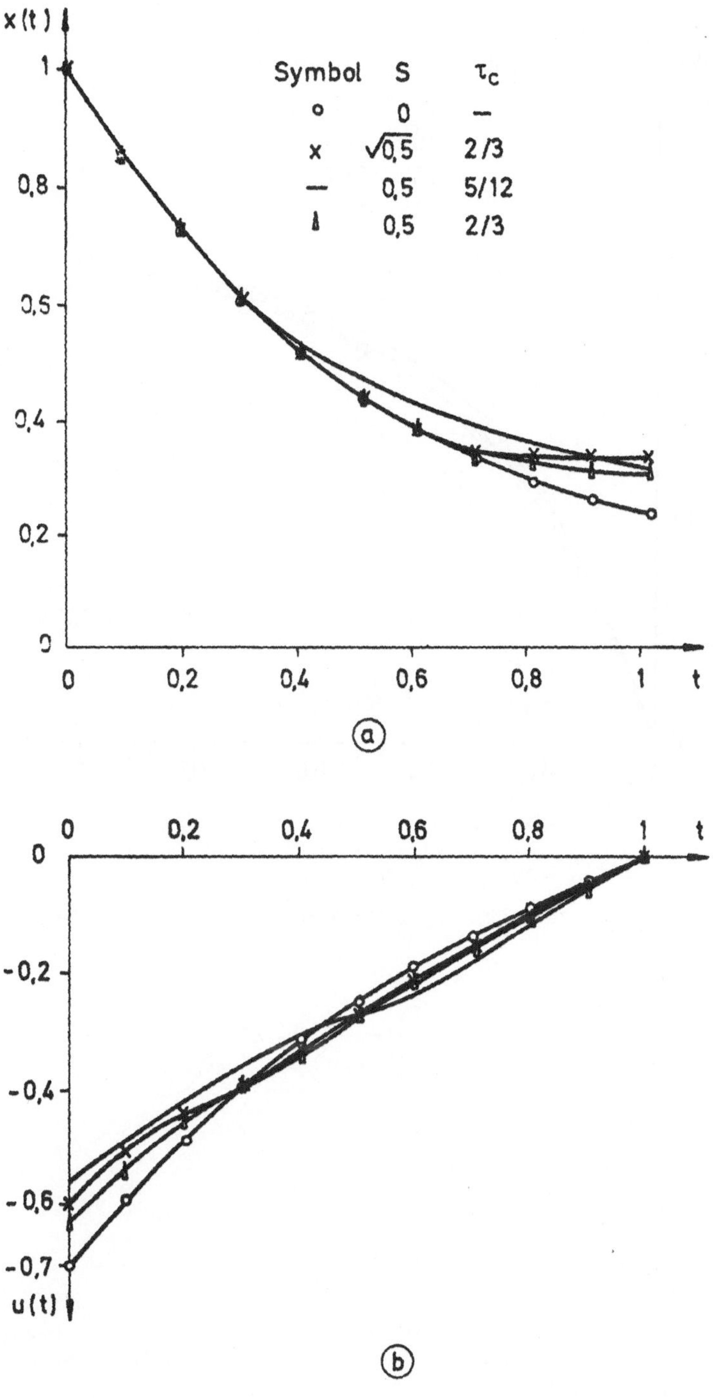

Fig. 8.7. Results of Example 8.9 $(m = 100)$

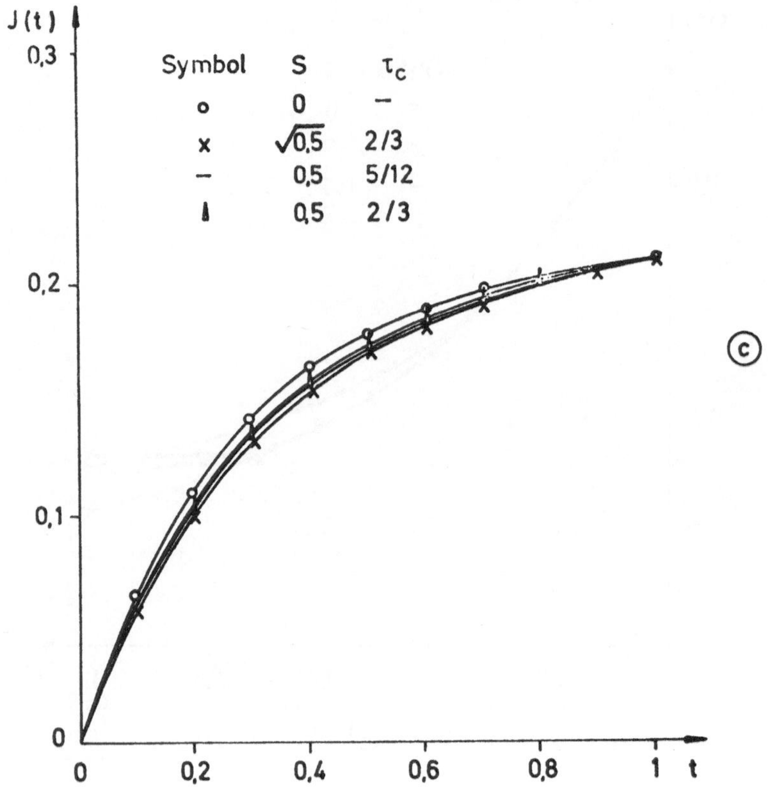

Fig. 8.7. Results of Example 8.9 (m = 100) contd.

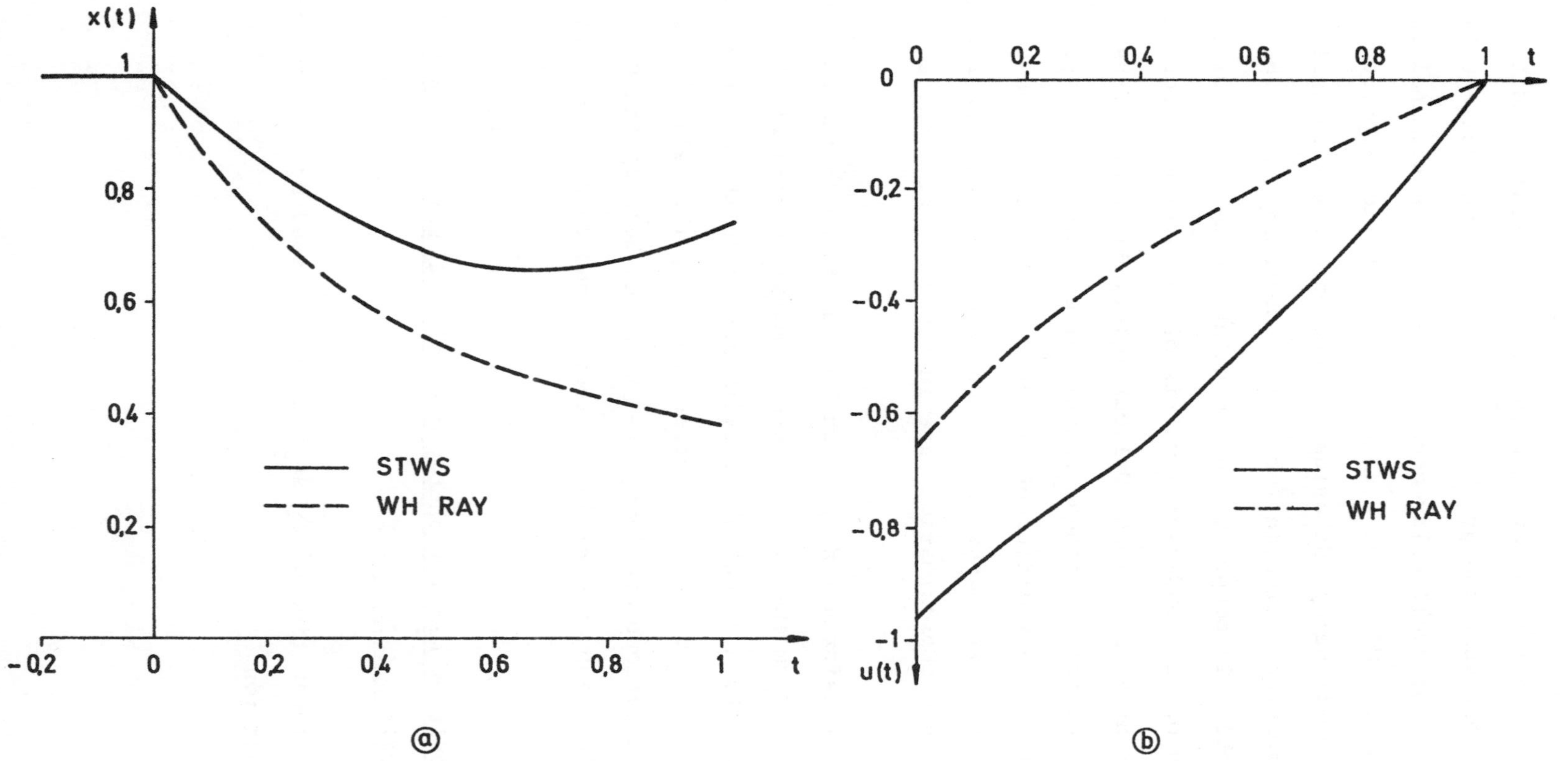

Fig. 8.8. Results of Example 8.10

It is observed that this approach meets the terminal conditions at the cost of increased energy. The conventional synthesis [C23] which ignores the small delay does not satisfy the terminal conditions. Sannuti [C44] gave a singular perturbation approach to the case of small delay. Rekasius and Lawrence [C40] suggest an iterative technique suitable for systems containing small delays. The main problem is computational in nature. Often we require either the solution or the fundamental matrix of the systems described by delay-differential equations. Delfour [C16], Banks and Burns [C7] approximate the related linear functional equations by a system of higher order ordinary differential equations and establish convergence in the context. Banks [C4] suggests an averaging approximation technique. These approximations are in the sense of least squares.

The Riccati transformation approach however fails in the case of problems of minimum energy with zero terminal constraint. Therefore we resort to a direct sensitivity method in the solution of problems as detailed in this section. That is, the related state and costate equations are integrated with the help of the PCBF techniques. In this approach we do not require the delay to be small.

8.11. Synthesis of minimum energy control with zero terminal constraints [W36,W37a]

Let a linear time-delay system be expressed by a delay differential equation

$$\left.\begin{aligned} &\dot{\underline{x}}(t) = \underline{A}\ \underline{x}(t) + \underline{L}\ \underline{x}(t-\tau_s) + \underline{B}\ \underline{u}(t), \\ &\text{with initial function} \\ &\underline{x}(t) = x_b(t),\ -\tau_s \le t \le 0, \end{aligned}\right\} \tag{8.34}$$

where $\underline{x}$, $\underline{u}$, $\underline{A}$, $\underline{L}$, $\underline{B}$, and τ_s are respectively the state vector, control vector, nxn, nxr constant matrices and a positive scalar time delay parameter.

The synthesis problem of minimum energy control with terminal constraints is given as follows [C25]. For a system expressed by (8.34), synthesize an optimal control u*(t) such that it satisfies the terminal constraint

$$\underline{x}(T) = \underline{x}^f,$$

and minimizes the energy consumption

$$J = \frac{1}{2}\int_0^T \underline{u}^T(t)\ \underline{R}\ \underline{u}(t)\ dt, \tag{8.35}$$

where T, $\underline{x}^f$ and $\underline{R}$ are respectively a given final time, a given desired final state and a positive definite matrix.

When the delay time τ_s is exactly known, this synthesis problem is reduced to the problem of solving the following two point boundary value problem (TPBVP) by applying the Maximum Principle derived by Kharatishvili [C25]

$$\dot{\underline{x}}(t) = \underline{A}\ \underline{x}(t) + \underline{L}\ \underline{x}(t-\tau_s) - \underline{B}\ \underline{R}^{-1}\ \underline{B}^T\ \underline{p}(t), \tag{8.36a}$$

$$\dot{\underline{p}}(t) = -\underline{A}^T \underline{p}(t) - \underline{L}^T \underline{p}(t+\tau_s), \tag{8.36b}$$

and

$$\underline{x}(t) = \underline{x}_b(t); \; -\tau_s \leq t \leq 0; \; \underline{x}(T) = \underline{x}^f, \tag{8.36c}$$

$$\underline{p}(t) = 0, \; T < t < T + \tau_s, \tag{8.36d}$$

and the optimal control $\underline{u}^*(t)$ is given by

$$\underline{u}^*(t) = -\underline{R}^{-1} \underline{B}^T \underline{p}(t), \tag{8.36e}$$

where p(t) is the adjoint vector.

The TPBVP (8.36) is not easy to solve in practice, since (8.36a) is of retarded type, while (8.36b) is of advanced type. Equations (8.36a) and (8.36b) can be solved by sensitivity method [C29]. The method is as follows:

Assuming some arbitrary value of $\underline{p}(T)$, (8.36b) is solved backward and $\underline{p}(t)$ thus obtained is then used in (8.36a) to obtain $\underline{x}(T)$ starting with given initial function $\underline{x}_b(t)$. The $\underline{x}(T)$ thus obtained would be different from optimal value $\underline{x}^*(T) = \underline{x}^f$ since $\underline{p}(T)$ is arbitrary. If the choice of $\underline{p}(T)$ happens to be optimal, $\underline{x}(T)$ so found would be $\underline{x}^f$. A change $\Delta\underline{x}(T)$ about $\underline{x}^f$ due to small change $\Delta\underline{p}(T)$ about optimal value $\underline{p}^*(T)$ can be written as

$$\Delta\underline{x}(T) = \frac{\Delta\underline{x}(T)}{\Delta\underline{p}(T)} \; \underline{p}(T), \tag{8.37}$$

where

$$\Delta\underline{x}(T) = \underline{x}^f - \underline{x}(T), \tag{8.38}$$

if

$$\underline{x}^f = 0, \quad \underline{x}(T) = -\left[\frac{\Delta \underline{x}(t)}{\Delta \underline{p}(T)}\right]_{t=T} \Delta \underline{p}(T). \tag{8.39}$$

The change in $\underline{x}(T)$ is related with the sensitivity coefficient $S_c = \left[\frac{\Delta \underline{x}(t)}{\Delta \underline{p}(t)}\right]_{t=T}$ of $\underline{x}(t)$ and $\Delta \underline{p}(T)$ [C36].

Direct sensitivity method

Levine's iterative steepest-descent method [C30] is modified by Paul et al.[C36] to a direct sensitivity method. The method is applied here. Let us consider the following sensitivity equations of (8.36a-b) for a small change in p(T).

$$\dot{\underline{y}}(t) = \underline{A}\,\underline{y}(t) - \underline{L}\,\underline{y}(t-\tau_s) - \underline{B}\,\underline{R}^{-1}\,\underline{B}^T\,\underline{z}(t), \tag{8.40a}$$

$$\dot{\underline{z}}(t) = -\underline{A}^T\,\underline{z}(t) - \underline{L}^T\,\underline{z}(t+\tau_s), \tag{8.40b}$$

where $\frac{\partial \underline{x}(t)}{\partial \underline{p}(T)} = \underline{y}(t)$, $\frac{\partial \underline{p}(t)}{\partial \underline{p}(T)} = \underline{z}(t)$, $\frac{\partial \underline{x}(0)}{\partial \underline{p}(T)} = 0$ since $\underline{x}(0)$ is independent of $\underline{p}(T)$,

$\frac{\partial \underline{x}(t-\tau_s)}{\partial \underline{p}(T)} = 0$, $t \in [-\tau_s, 0]$ since the change in $x(t-\tau_s)$ is zero since the initial function is a constant,

$\frac{\partial \underline{p}(t+\tau_s)}{\partial \underline{p}(T)} = 0$, $t > T$ since $\underline{p}(t) = 0$ for $t > T$.

Equation (8.40b) may be solved by backward integration and (8.40a) by forward integration using the above conditions. Then $\Delta\underline{p}(T)$ may be calculated from the following equation

$$\left[\frac{\partial \underline{x}(t)}{\partial \underline{p}(t)}\right]_{t=T} \underline{p}(T) = \underline{x}^f - \underline{x}(T). \qquad (8.40)$$

At first $\underline{x}(t_f)$ is calculated using (8.36a-b) assuming some $\underline{p}(T)$. In the next step $\underline{p}(T)$ is modified using $\Delta\underline{p}(T)$ from (8.40a-c). The modified value of $\underline{p}(T)$ is used in (8.36a-b) to get optimal values of $\underline{x}^*(t)$ and $\underline{p}^*(t)$. There is no further iteration as far as linear problems are concerned. The direct sensitivity method gives the solution in two steps.

The computational algorithm may be written as follows:

1. Assume $\underline{p}(T)$ arbitrarily and solve (8.36b) backward.
2. Using $\underline{p}(T)$ obtained in step (1), solve (8.36a) to obtain $\underline{x}(T)$.
3. Change the value of $\underline{p}(T)$ by an amount $\Delta\underline{p}(T)$ obtained by sensitivity method.
4. Repeat steps 1 to 2 only once with the new value of $\underline{p}(T)$.

Solution of adjoint and state equations

Following the algorithm above, the adjoint and state equations (8.36b) and (8.36a) are solved by single term PCBF method. To normalise the interval $[0,1/m)$ to $[0,1)$ let $\hat{t} = m\,t$ in equations (8.36a) and (8.36b). Let us express $\dot{\underline{x}}(\hat{t})$, $\underline{x}(\hat{t})$, $\underline{x}(\hat{t}-m\tau_s)$, $\dot{\underline{p}}(\hat{t})$, $\underline{p}(\hat{t}+m\tau_s)$ and $\underline{x}_b(\hat{t})$ in single PCBF as follows:

$$\left.\begin{aligned}
&\dot{\underline{x}}(\hat{t}) = \underline{V}_x^{(j)}\ \theta_1(\hat{t}), \quad \dot{\underline{p}}(\hat{t}) = \underline{V}_p^{(j)}\ \theta_1(\hat{t}),\\
&\underline{x}(\hat{t}) = \underline{Y}_x^{(j)}\ \theta_1(\hat{t}), \quad \underline{p}(\hat{t}) = \underline{Y}_p^{(j)}\ \theta_1(\hat{t}),\\
&\underline{x}(\hat{t} \pm m\tau_s) = \underline{Y}_x^{(j\pm\sigma)}\ \theta_1(\hat{t}),\\
&\underline{p}(\hat{t} \pm m\tau_s) = \underline{Y}_p^{(j\pm\sigma)}\ \theta_1(\hat{t}),\\
&\text{and}\\
&\underline{x}_o(\hat{t}) = \underline{Y}_x^{(i)}\ \theta_1(\hat{t}) \text{ for } i = -(m\tau_s-1)\ \ldots\ -1, 0\,.
\end{aligned}\right\} \quad (8.41)$$

Equation (8.36b) is solved for $\underline{p}(t)$ by backward integration as

$$\begin{aligned}
\underline{V}_p^{(j)} &= \left[\underline{I} - \frac{\underline{A}^T}{2m}\right]^{-1} \left[\frac{\underline{A}^T}{m}\ \underline{p}(t_j) + \frac{\underline{L}^T}{m}\ \underline{Y}_p^{(j+\sigma)}\right],\\
\underline{Y}_p^{(j)} &= \frac{1}{2}\ \underline{V}_p^{(j)} + \underline{p}(t_j),\\
\underline{p}(t_{j-1}) &= \underline{V}_p^{(j)} + \underline{p}(t_j),
\end{aligned} \quad (8.42)$$

where $j = mT\ \ldots\ 3,\ 2,\ 1.$

$\underline{p}(t_j)$ represents the discrete time values of $\underline{p}(t)$, $\underline{p}(t) = 0$, $t > T$ and $\sigma = m\tau_s$ the number of segements delay or advance.

Using the $\underline{Y}_p$ values obtained from (8.42), (8.36a) is solved by forward integration using (8.41),

$$\left.\begin{aligned}
\underline{V}_x^{(j)} &= \left[\underline{I} - \frac{\underline{A}}{2m}\right]^{-1} \left[\frac{\underline{A}}{m}\,\underline{x}(t_{j-1}) + \frac{\underline{L}}{m}\,\underline{Y}_x^{(j-\sigma)} - \frac{\underline{B}\,\underline{R}^{-1}\,\underline{B}^T}{m}\,\underline{Y}_p^{(j)}\right], \\
\underline{Y}_x^{(j)} &= \frac{1}{2}\,\underline{V}_x^{(j)} + \underline{x}(t_{j-1}), \\
\underline{x}(t_j) &= \underline{V}_x^{(j)} + \underline{x}(t_{j-1}),\quad \underline{x}(t_o) = \underline{x}_b(0), \\
j &= 1,\ 2,\ \ldots\ mT.
\end{aligned}\right\} \quad (8.43)$$

and $\underline{x}(t_j)$ represents discrete time values of the state. Equations (8.36a) and (8.36b) are solved using the given algorithm.

8.12. Illustrative examples contd.

EXAMPLE 8.11:

Consider the scalar system [C22,C23]

$$\left.\begin{aligned}
\dot{x}(t) &= ax(t) + bx(t-\tau_s) + u(t), \\
x(t) &= 1.0 \qquad -\tau_s \leq t \leq 0,
\end{aligned}\right.$$

with the performance index

$$J = \frac{1}{2}\int_0^1 u^2(t)\,dt. \qquad (8.44)$$

The state and costate equations are

$$\dot{x}(t) = ax(t) + bx(t-\tau_s) - p(t)\ , \qquad (8.45a)$$

$$\dot{p}(t) = -ap(t) - bp(t+\tau_s)\ ,\quad p(t) = 0,\ t > T \qquad (8.45b)$$

Now, equations (8.45) are transformed into single term PCBF version as a recursion:

$$\left.\begin{aligned}
V_p^{(j)} &= \left[1-\frac{a}{2m}\right]^{-1}\left[-\frac{a}{m}\,p(t_j) + \frac{b}{m}Y_p^{(j+\sigma)}\right],\\
Y_p^{(j)} &= 0,\quad j > mT,\qquad j = mT \ldots 1 \text{ and } \sigma = m\tau_s.\\
V_x^{(j)} &= \left[1-\frac{a}{2m}\right]^{-1}\left[\frac{a}{m}\,x(t_{j-1}) + \frac{b}{m}\,Y_x^{(j-\sigma)} - Y_p^{(j)}\right],\\
Y_x^{(j)} &= 1.0,\quad (\sigma-1) \le j \le 0\\
x(t_o) &= 1.0 \text{ and } j = 1,2,\ldots,mT\,.
\end{aligned}\right\} \tag{8.46a}$$

The sensitivity equations are

$$\dot{y}(t) = ay(t) + by(t-\tau_s) - z(t)\,, \tag{8.46b}$$

$$\dot{z}(t) = -az(t) - bz(t+\tau_s)\,, \tag{8.46c}$$

with

$$y(0) = 0.0\,,\ y(t) = 0\,,\ t \in [-\tau_s,0]$$

$$z(T) = 1.0,\ \text{and } z(t) = 0\quad t > T\,.$$

Y_p, $p(t)$, Y_x and $x(t)$ are calculated as given by (8.43) and (8.42) at every step. Assuming $p(T)$ equations (8.46a) are solved to get $x(T)$. Then $\Delta p(T)$ is obtained by solving sensitivity equations (8.46a) and (8.46b). Modifying the value of $p(T)$ using $\Delta p(T)$ equations (8.46a) are solved to get optimal values. The obtained results for $a=2$, $B=0$; $a=1$, $b=1$, $\tau_s=0.1$; $a=-2.0$, $b=0$, and $a=-1.0$, $b=-1.0$, $\tau_s=0.1$ are shown in Figs. 8.9a-d with comparison.

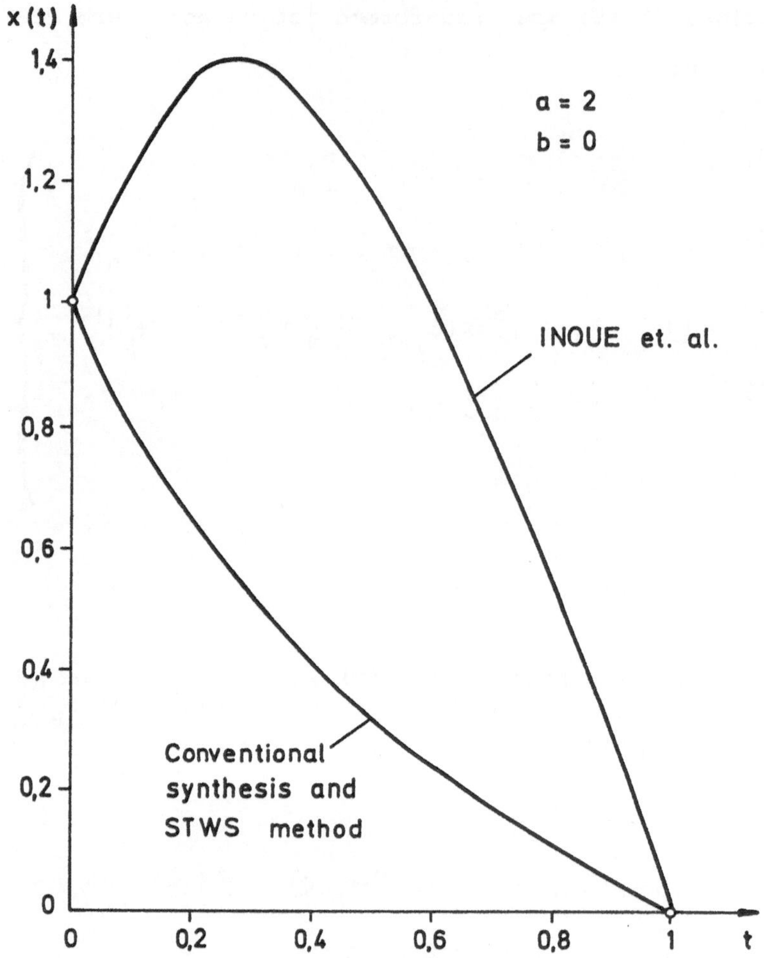

Fig. 8.9 a. Results of Example 8.11

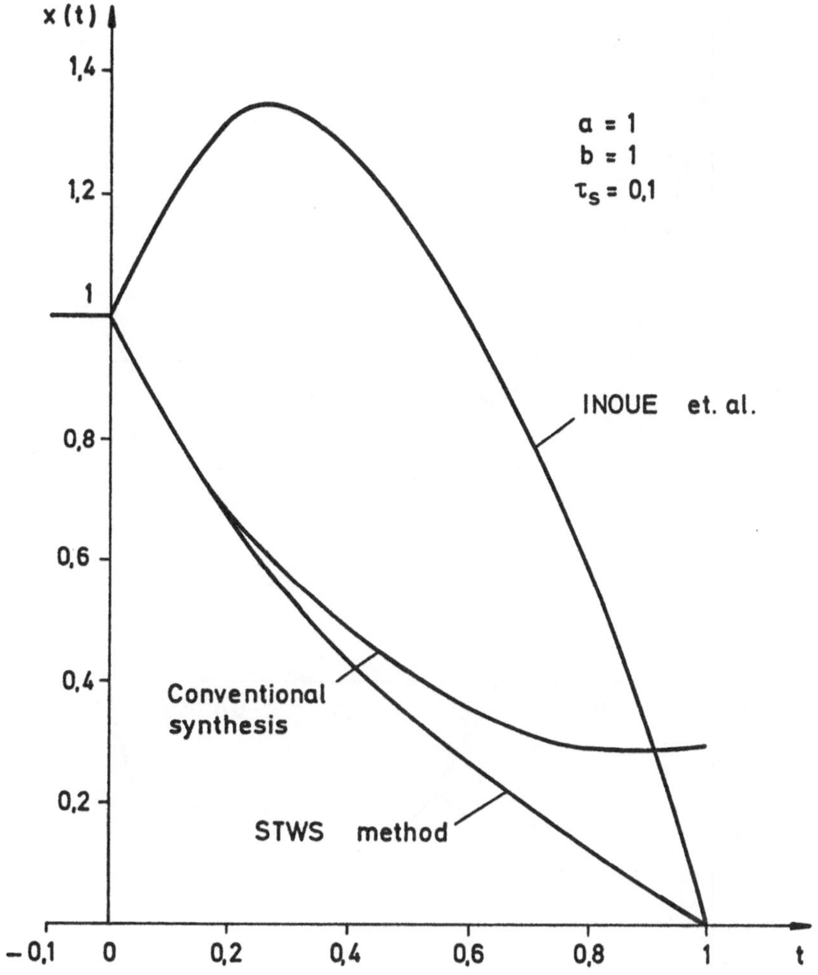

Fig. 8.9 b. Results of Example 8.11 contd.

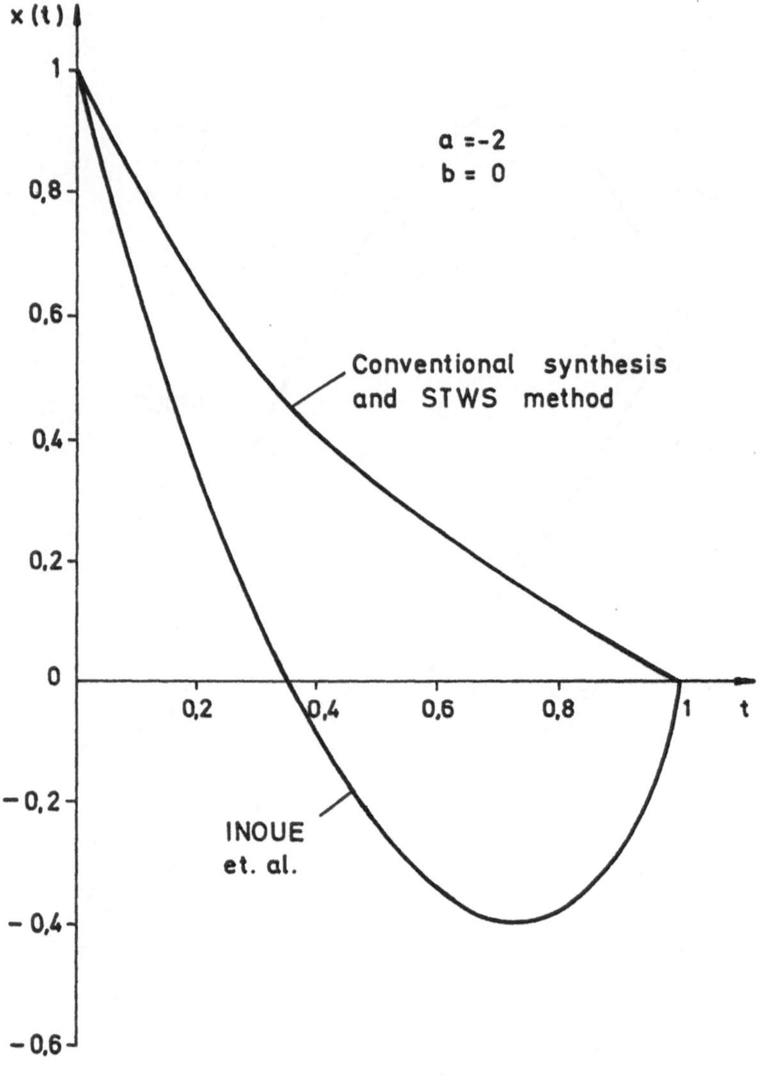

Fig. 8.9 c. Results of Example 8.11 contd.

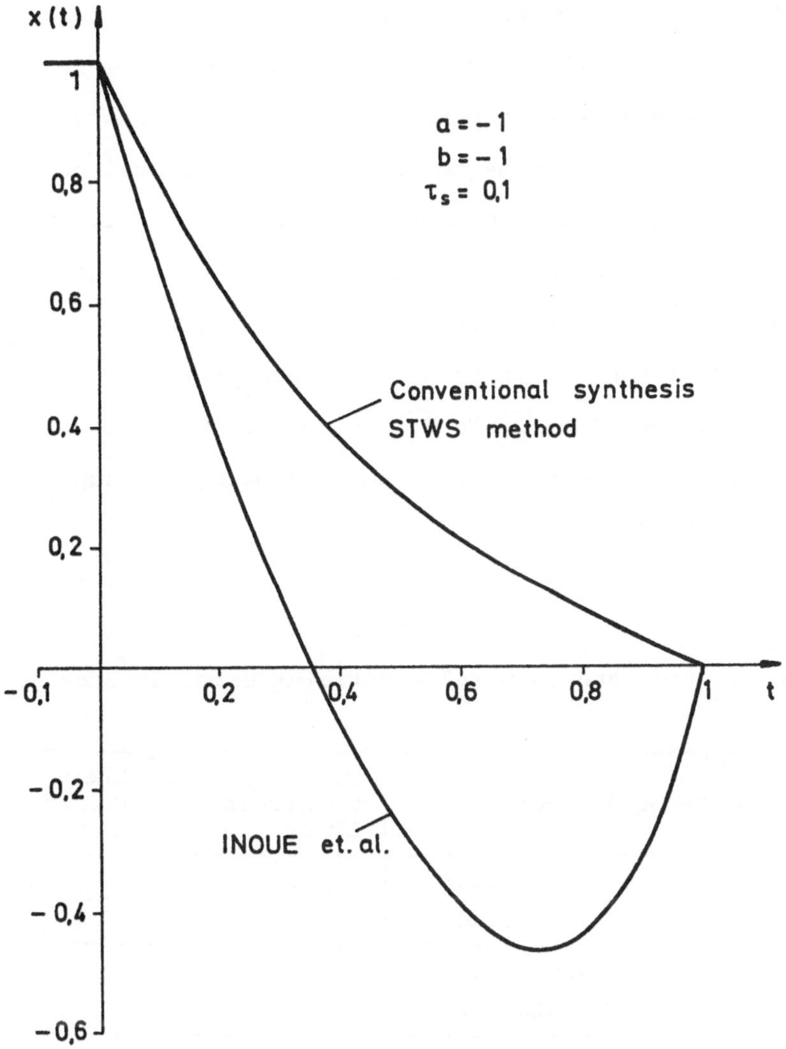

Fig. 8.9 d. Results of Example 8.11 contd.

Table 8.4a presents a comparison of the results obtained by Inoue et al. with those by the present method. As seen from the table Inoue et al's sensitivity synthesis consumes more energy and the conventional synthesis leads to considerable terminal constraint error. The present method consumes less energy and gives negligible terminal constraint error including fully the effect of the delay. Unlike Inoue et al's method the present one may be applied to systems having large delays as well and does not increase the dimensions of the problem. Table 8.4b compares the terminal constraint error for different magnitudes of delay showing the superiority of the present method.

TABLE 8.4a

Energy consumption and terminal constraint error in example 8.11

System	Delay	Method	Terminal Constraint Error	Energy Consumption
a=1.0 b=1.0	0	conventional synthesis	0.2933	2.0373
		sensitivity synthesis	0.0078	5.4181
		STWS	0.0	2.2381
a=-1.0 b=-1.0	0	conventional synthesis	-0.0356	0.0373
	0.1	sensitivity synthesis	-0.0122	1.4181
	0.1	STWS	0.0	0.0209

TABLE 8.4b

Terminal constraint error for different delays in example 8.11

Actual System	Method	Terminal Constraint Error			
		Delay			
		0.05	0.10	0.15	0.20
$a=1.0$	C.S.	0.1686	0.2933	0.3900	0.4674
$b=1.0$	S.S.	0.0030	0.0078	0.0123	0.0158
$\gamma_x(t)=1.0$	STWS	0.0	0.0	0.0001	0.0
$a=-1.0$	C.S.	-0.0175	-0.0356	-0.0547	-0.0752
$b=-1.0$	S.S	-0.0030	-0.0122	-0.0277	-0.0522
$\gamma_x(t)=1.0$	STWS	0.0	0.0	-0.0001	0.0

C.S.: Conventional Synthesis
S.S.: Sensitivity Synthesis

EXAMPLE 8.12:

Consider the system with large delay [C29]

$$\dot{x}(t) = -x(t) - 0.5\,x(t-1) + u(t), \tag{8.47}$$

$$x(0) = 1.0,$$

$$x(10) = 0.0, \quad x(t) = 0.0, \quad t < 0,$$

with the performance index $J = \frac{1}{2}\int_0^{10} u^2(t)\,dt$. The results are shown in Fig. 8.10.

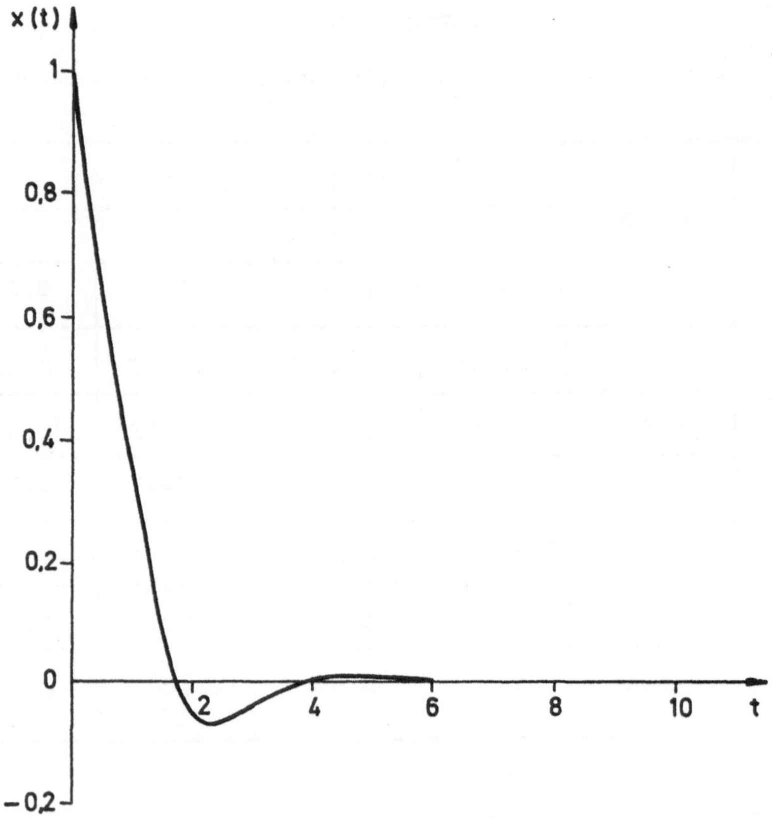

Fig. 8.10. Results of Example 8.12

IX

SOLUTION OF PARTIAL DIFFERENTIAL EQUATIONS (PDE) [W55]

9.1.

The use of multidimensional PCBF series and the various operational matrices for integration in several dimensions in obtaining piecewise constant solutions of partial differential equations (PDE) will now be discussed. Three methods of computation with increasing simplicity, all leading to the same results will be presented. We consider first and second order PDE in two dimensions x and t and use two dimensional BPF as the basis for illustration. Further, for notational simplicity we drop the suffix β from all multidimensional terms. That is, for instance, $\underline{E}_{\beta_x}$, $\underline{E}_{\beta_t}$, $\underline{E}_{\beta_{xt}}$ etc. are written simply as $\underline{E}_x$, $\underline{E}_t$, $\underline{E}_{xt}$ etc. respectively.

The following notations will be followed in the sequel. Let u(x,t) be a function of two variables x and t. Let there be m segments over [0,1) along the x axis and n segments over [0,1) along the t axis. Let the following block pulse series expansions be common to all problems for the sake of notational convenience

$$\left.\begin{aligned}
u(x,t) &\equiv \underline{\underline{u}}^T\underline{\beta}(x,t) = [u_{1,1}\cdots u_{1,n}u_{2,1}\cdots u_{2,n}\\
&\qquad\cdots u_{m,1}\cdots u_{m,n}]\underline{\beta}(x,t),\\
u(0,t) = f(t) &\equiv \underline{\underline{f}}^T\underline{\beta}(x,t) = [f_{1,1}\cdots f_{1,n}f_{2,1}\cdots f_{2,n}\\
&\qquad\cdots f_{m,1}\cdots f_{m,n}]\underline{\beta}(x,t),\\
u(x,0) = g(x) &\equiv \underline{\underline{g}}^T\underline{\beta}(x,t) = [g_{1,1}\cdots g_{1,n}g_{2,1}\cdots g_{2,n}\\
&\qquad\cdots g_{m,1}\cdots g_{m,n}]\underline{\beta}(x,t),\\
\frac{\partial u}{\partial x}(0,t) = h(t) &\equiv \underline{\underline{h}}^T\underline{\beta}(x,t) = [h_{1,1}\cdots h_{1,n}h_{2,1}\cdots h_{2,n}\\
&\qquad\cdots h_{m,1}\cdots h_{m,n}]\underline{\beta}(x,t),\\
\frac{\partial u}{\partial t}(x,0) = q(x) &\equiv \underline{\underline{q}}^T\underline{\beta}(x,t) = [q_{1,1}\cdots q_{1,n}q_{2,1}\cdots q_{2,n}\\
&\qquad\cdots q_{m,1}\cdots q_{m,n}]\underline{\beta}(x,t).
\end{aligned}\right\} \quad (9.1)$$

In the above expressions it should be noted that $f_{i,j}$, $g_{i,j}$, $h_{i,j}$ and $q_{i,j}$ all reduce to the coefficients of BPFs of a single variable since they are all functions of a single variable. Further it should be noted that

(i) $f_{i,j} = f_{1,j}$ and $h_{i,j} = h_{1,j}$ for $i=2,3,\ldots,m$

(ii) $g_{i,j} = g_{i,1}$ and $q_{i,j} = q_{i,1}$ for $j=2,3,\ldots,n$

<u>Example 9.1.:</u>

Consider the equation

$$\alpha\frac{\partial u}{\partial x}(x,t) + \frac{\partial u}{\partial t}(x,t) = 0\,, \tag{9.2}$$

with $u(0,t)=f(t)$ and $u(x,0)=g(x)$ where α is a real constant.

Let $(\partial u/\partial x)(x,t)$ and $(\partial u/\partial t)(x,t)$ be expanded in terms of BPF series as

$$\frac{\partial u}{\partial x}(x,t) \equiv \underline{\underline{a}}^T\underline{\beta}(x,t) = [a_{1,1}\ldots a_{1,n}a_{2,1}\ldots a_{2,n}\ldots a_{m,1}\ldots a_{m,n}]\underline{\beta}(x,t),$$

$$\frac{\partial u}{\partial t}(x,t) \equiv \underline{\underline{b}}^T\underline{\beta}(x,t) = [b_{1,1}\ldots b_{1,n}b_{2,1}\ldots b_{2,n}\ldots b_{m,1}\ldots b_{m,n}]\underline{\beta}(x,t). \tag{9.3}$$

Equation (9.2) can be rewritten in terms of the two-dimensional BPF series expansion as:

$$\alpha\underline{\underline{a}}^T\underline{\beta}(x,t) + \underline{\underline{b}}^T\underline{\beta}(x,t) = 0\,. \tag{9.4}$$

Method (a): "Kronecker product" formula

Integrating eqn.(9.2) with respect to both x and t we get

$$\alpha\int_0^x\int_0^t \frac{\partial u}{\partial x}\,dt\,dx + \int_0^t\int_0^x \frac{\partial u}{\partial t}\,dx\,dt = 0\,, \tag{9.5}$$

or

$$\alpha \int_0^t [u(x,t) - u(0,t)]dt + \int_0^x [u(x,t) - u(x,0)]dx = 0 . \quad (9.6)$$

Employing the operational matrices for integration eqn.(9.6) can be rewritten in terms of the BPF series expansion as

$$(\alpha \underline{\underline{u}}^T \underline{E}_t - \alpha \underline{\underline{f}}^T \underline{E}_t + \underline{\underline{u}}^T \underline{E}_x - \underline{\underline{g}}^T \underline{E}_x) \underline{\beta}(x,t) = 0 . \quad (9.7)$$

Equation (9.7) leads to

$$\underline{\underline{u}}^T = (\alpha \underline{\underline{f}}^T \underline{E}_t + \underline{\underline{g}}^T \underline{E}_x)(\alpha \underline{E}_t + \underline{E}_x)^{-1}, \quad (9.8)$$

where $\underline{E}_x = (\underline{E}_m \otimes \underline{I}_n)$ and $\underline{E}_t (\underline{I}_m \otimes \underline{E}_n)$ as defined in § 2.1.7. Equation (9.8) involves the addition, inversion and multiplication of matrices of order (mnxmn). Further $(\alpha \underline{E}_t + \underline{E}_x)^{-1}$ does not exist for $\alpha = -n/m$.

This method gives us a solution which is piecewise constant, unique and optimal in the sense of least squared error but does not suggest any way of determining the accuracy and stability of the solution.

Method (b): Recurrence formula

Here we develop a method, to obtain the solution $\underline{\underline{u}}^T$ which will avoid heavy matrix algebra and lead to a simple recurrence formula. We have from eqn. (9.3)

$$\int_0^x \frac{\partial u}{\partial x}(x,t)dx = \int_0^x \underline{\underline{a}}^T \underline{\beta}(x,t)dx , \quad (9.9a)$$

or

$$\underline{\underline{a}}^T = [\underline{\underline{u}}^T - \underline{\underline{f}}^T] \underline{E}_x^{-1} , \quad (9.9b)$$

and similarly

$$\underline{\underline{b}}^T = [\underline{\underline{u}}^T - \underline{\underline{g}}^T] \underline{E}_t^{-1} . \quad (9.9c)$$

From eqn. (9.4) we have

$$\alpha \underline{\underline{a}}^T = -\underline{\underline{b}}^T . \tag{9.10a}$$

Comparing the like columns on either side we get

$$\alpha a_{i,j} = -b_{i,j} . \tag{9.10b}$$

This means

$$\{\alpha[\underline{\underline{u}}^T - \underline{\underline{f}}^T]\underline{E}_x^{-1}\}_{i,j} = -\{[\underline{\underline{u}}^T - \underline{\underline{g}}^T]\underline{E}_t^{-1}\}_{i,j} . \tag{9.10c}$$

We have $\underline{E}_x^{-1} = (\underline{E}_m^{-1} \otimes \underline{I}_n)$ and $\underline{E}_t^{-1} = (\underline{I}_m \otimes \underline{E}_m^{-1})$. The use of $\underline{E}_x^{-1}$ and $\underline{E}_t^{-1}$ in eqn.(9.10c) leads to a general formula for $u_{i,j}$ as

$$(n+\alpha m)u_{i,j} = -2m\alpha \sum_{k=1}^{i-1} (-1)^{i-k} u_{k,j} - m\alpha(-1)^i f_{i,j} - 2n \sum_{i=1}^{j-1} (-1)^{j-1} u_{i.1} - n(-1)^j g_{i,j} , \tag{9.11}$$

for $1 \leq i \leq m$ and $1 \leq j \leq n$ with the following conditions:

(i) when i=1, the first term on the right-hand side is zero
(ii) when j=1, the third term on the right-hand side is zero
(iii) when i=j=1, the both (i) and (ii) jointly hold.

Use of eqn. (9.11) is clearly simpler and more convenient in comparison to the use of eqn.(9.8) to obtain a solution for $\underline{\underline{u}}^T$. However the situation $\alpha=-n/m$ should be avoided. Equation (9.11) also does not throw any light on the accuracy and stability of the solution.

Method (c): Recursion by single block extension

The region of solution under consideration is shown in Fig. 9.1a. If we scale the x and t axes such that $mx=\hat{x}$ and $nt=\hat{t}$ with $\hat{x}$ and $\hat{t}$ as our new coordinate axes then we seek a solution of $u(\hat{x},\hat{t})$ in a region

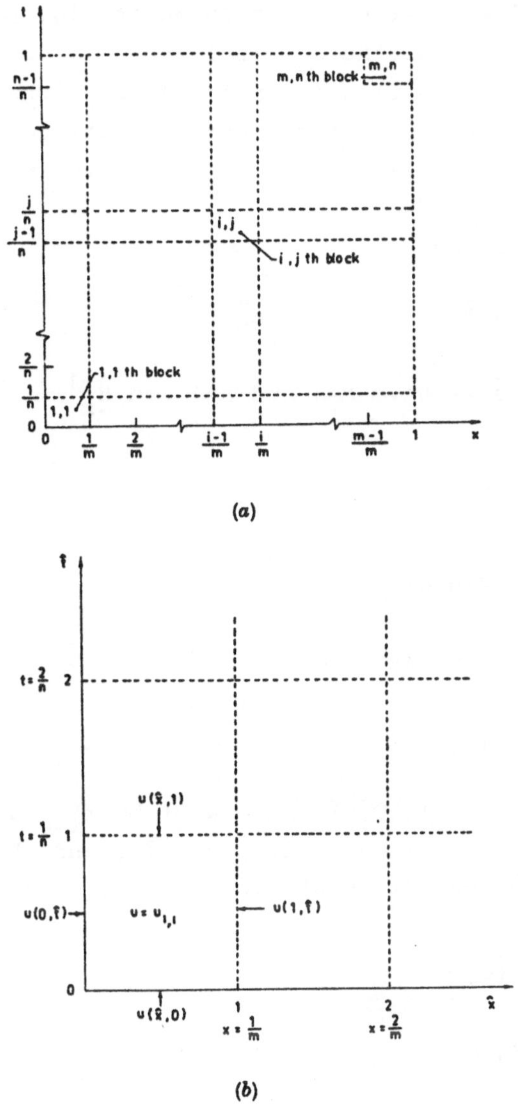

Fig. 9.1. (a) Region of solution of u(x,t) under consideration.
(b) The first block of the region under consideration after scaling

$\hat{x} \in (0,m]$ and $\hat{t} \in (0,n]$. The first block is shown in Fig. 9.1b. Here we solve for u which is $u_{1,1}$ of the original system. Further we obtain the boundary conditions $u(1,\hat{t})$ and $u(\hat{x},1)$. These boundary conditions become initial conditions for continuing the solution to obtain $u_{2,1}$ and $u_{1,2}$. In this fashion we obtain the solution in the whole region.

Equation (9.2) now becomes

$$\gamma\frac{\partial u}{\partial \hat{x}}(\hat{x},\hat{t}) = -\frac{\partial u}{\partial \hat{t}}(\hat{x},\hat{t}), \text{ where } \gamma = \alpha\frac{m}{n} . \tag{9.12}$$

Under these conditions we have m=1 and n=1 and $\underline{E}_x^{-1} = \underline{E}_t^{-1} = 2$, leading to

$$\begin{gathered} \underline{\underline{u}}^T = u,\ \underline{\underline{a}}^T = a,\ \underline{\underline{b}}^T = b,\ \underline{\underline{f}}^T = f,\ \underline{\underline{g}}^T = g, \\ \text{and} \\ \underline{\underline{a}} = 2[u-f],\quad \underline{\underline{b}} = 2[u-g]. \end{gathered} \tag{9.13}$$

Since $\gamma\underline{\underline{a}}=-\underline{\underline{b}}$ or $\gamma a=-b$, we have,

$$u = \frac{1}{1+\gamma}[\gamma f+g] = u_{1,1} . \tag{9.14}$$

Let the BPF components of $(u(0,\hat{t})$ and $u(\hat{x},0)$ be $f^*_{i=o}$ and $g^*_{j=o}$ and the BPF components of $u(1,\hat{t})$ and $u(\hat{x},1)$ be $f^*_{i=1}$ and $g^*_{j=1}$ respectively. We can show that (see Appendix 9.3)

$$\begin{aligned} f^*_{i=1} &= 2u_{1,1} - f^*_{i=o} , \\ g^*_{j=1} &= 2u_{1,1} - g^*_{j=o} . \end{aligned} \tag{9.15}$$

Equations (9.14) and (9.15) can be rewritten as

$$\begin{bmatrix} u \\ f^*_{i=1} \\ g^*_{j=1} \end{bmatrix} = \begin{bmatrix} 0 & \frac{\gamma}{1+\gamma} & \frac{\gamma}{1+\gamma} \\ 0 & -1 & 0 \\ 0 & 0 & -1 \end{bmatrix} \begin{bmatrix} 0 \\ f^*_{i=o} \\ g^*_{j=o} \end{bmatrix} + \begin{bmatrix} 0 & 0 & 0 \\ 2 & 0 & 0 \\ 2 & 0 & 0 \end{bmatrix} \begin{bmatrix} u \\ f^*_{i=1} \\ g^*_{j=1} \end{bmatrix} . \tag{9.16}$$

Equation (9.16) describes the situation shown in Fig. 9.1b. In order to generalize this into the original situation shown in Fig. 9.1a, we return to the double subscribt notation and rewrite eqn.(9.16) as

$$\begin{bmatrix} u_{i,j} \\ f^*_{i,j} \\ g^*_{i,j} \end{bmatrix} = \begin{bmatrix} 0 & \frac{\gamma}{1+\gamma} & \frac{1}{1+\gamma} \\ 0 & -1 & 0 \\ 0 & 0 & -1 \end{bmatrix} \begin{bmatrix} 0 \\ f^*_{i-1,j} \\ g^*_{i,j-1} \end{bmatrix} + \begin{bmatrix} 0 & 0 & 0 \\ 2 & 0 & 0 \\ 2 & 0 & 0 \end{bmatrix} \begin{bmatrix} u_{i,j} \\ f^*_{i,j} \\ g^*_{i,j} \end{bmatrix}.$$

On rearranging the above expression we obtain

$$\begin{bmatrix} u_{i,j} \\ f^*_{i,j} \\ g^*_{i,j} \end{bmatrix} = \begin{bmatrix} 0 & \frac{\gamma}{1+\gamma} & \frac{1}{1+\gamma} \\ 0 & -\frac{1-\gamma}{1+\gamma} & \frac{2}{1+\gamma} \\ 0 & \frac{2\gamma}{1+\gamma} & \frac{1-\gamma}{1+\gamma} \end{bmatrix} \begin{bmatrix} 0 \\ f^*_{i-1,j} \\ g^*_{i,j-1} \end{bmatrix}. \tag{9.17}$$

The recurrence equation is by far the simplest in the context of this problem.

The transition matrix is valid over the entire region $0 \leq x \leq 1$ and $0 \leq t \leq 1$ and has eigenvalues +1, 0, -1 which are independent of segmentation. This shows that the errors are bounded. As an illustration let us consider the following example:

$$\frac{\partial u}{\partial x}(x,t) + \frac{\partial u}{\partial t}(x,t) = 0 \text{ with } u(0,t) = e^{-t} \text{ and } u(x,0) = e^x. \tag{9.18}$$

The piecewise constant solution obtained with m=5 and n=5 is shown in Fig. 9.2.

The 25 values of $u_{i,j}$ obtained by the recurrence relation (9.11) and the single block extension formula (9.17) are compared with the piecewise constant values of the exact solution $u(x,t) = e^{x-t}$ in Table 9.1.

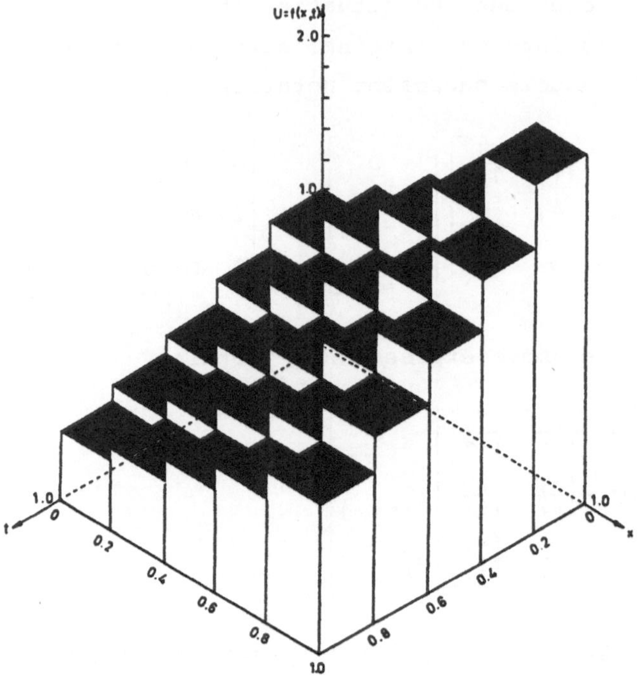

Fig. 9.2. Two-dimensional BPF solution of eqn.(9.8)

Example 9.2.

Consider the diffusion equation

$$\frac{\partial u}{\partial t}(x,t) = \alpha^2 \frac{\partial^2 u}{\partial x^2}(x,t),$$

with (9.19)

$$u(0,t) = f(t),\ u(x,0) = g(x) \text{ and } \frac{\partial u}{\partial x}(0,t) = h(t).$$

Let the BPF expansion series of $(\partial u/\partial t)(x,t)$ and $(\partial^2 u/\partial x^2)(x,t)$ in $0 \leq x \leq 1$, $0 \leq t \leq 1$, be

$$\frac{\partial u}{\partial t}(x,t) = \underline{\underline{a}}^T \underline{\beta}(x,t) \quad \text{and} \quad \frac{\partial^2 u}{\partial x^2}(x,t) = \underline{\underline{b}}^T \underline{\beta}(x,t). \tag{9.20}$$

Also let m and n be the number of segments over unit intervals on the x and t axes respectively.

Table 9.1. Solution of eqn.(9.18) via two-dimensional block-pulse functions. Piecewise constant values of $u_{i,j}$

i,j	Method (b)	Method (c)	Exact values
1,1	1.0066805	1.00668087	1.0033388
2,1	1.2295618	1.22956275	1.2254791
3,1	1.5017881	1.5017891	1.4968014
4,1	1.8342896	1.83429005	1.8282003
5,1	2.2404070	2.2304075	2.2329664
1,2	0.82419938	0.824199765	0.82146460
2,2	1.0066795	1.00668087	1.0033369
3,2	1.2295599	1.22956275	1.2254772
4,2	1.5017862	1.5017891	1.4968042
5,2	1.8342896	1.83429005	1.8281975
1,3	0.67479771	0.674797745	0.67255849
2,3	0.82419932	0.824199765	0.82146364
3,3	1.0066776	1.00668087	1.0033360
4,3	1.2295570	1.22956275	1.2254801
5,3	1.5017853	1.5017891	1.4968033
1,4	0.55247784	0.55247784	0.55064452
2,4	0.67479753	0.674797745	0.67255771
3,4	0.82419777	0.824199765	0.82146269
4,4	1.0066757	1.00668087	1.0033379
5,4	1.2295551	1.22956275	1.2254791
1,5	0.45233059	0.45069167	0.45082927
2,5	0.55247831	0.55083895	0.55065332
3,5	0.67479753	0.674797745	0.67255634
4,5	0.82419729	0.824199765	0.82146376
5,5	1.0066748	1.00668087	1.0033369

Method (a): 'Kronecker product' formula

Integrating eqn. (9.19) once with respect to t and twice with respect to x we get

$$\int_0^x\int_0^x\int_0^t \frac{\partial u}{\partial t}(x,t)\,dt\,dx^2 = \alpha^2\int_0^t\int_0^x\int_0^x \frac{\partial^2 u}{\partial x^2}dx^2dt\,,$$

or

$$\int_0^x\int_0^x [u(x,t)-u(x,0)]dx^2 = \alpha^2\int_0^t [u(x,t)-u(0,t) - \int_0^x \frac{\partial u}{\partial x}(0,t)dx]dt. \quad (9.21)$$

Expressing eqn.(9.21) in terms of BPFs and their operational matrices for integrations we have

$$[\underline{\underline{u}}^T-\underline{\underline{g}}^T]\underline{E}_x^2\underline{\beta}(x,t) = \alpha^2[\underline{\underline{u}}^T-\underline{\underline{f}}^T-\underline{\underline{h}}^T\underline{E}_x]\underline{E}_t\underline{\beta}(x,t). \quad (9.22)$$

Equation (9.22) leads to the solution of $\underline{\underline{u}}^T$ as

$$\underline{\underline{u}}^T = \{\underline{\underline{g}}^T\underline{E}_x^2-\alpha^2(\underline{\underline{f}}^T+\underline{\underline{h}}^T\underline{E}_x)\underline{E}_t\}(\underline{E}_x^2-\alpha^2\underline{E}_t)^{-1}. \quad (9.23)$$

To solve eqn.(9.23) m and n should be so chosen that $(\underline{E}_x^2-\alpha^2\underline{E}_t)^{-1}$ exists.

Method (b): Recurrence formula

Equation (9.19) can be written, in view of eqn.(9.20) as

$$\underline{\underline{a}}^T\underline{\beta}(x,t) = \alpha^2\underline{\underline{b}}^T\underline{\beta}(x,t), \quad (9.24a)$$

or

$$\underline{\underline{a}}^T = \alpha^2\underline{\underline{b}}^T. \quad (9.24b)$$

But

$$\underline{\underline{b}}^T = [\underline{\underline{u}}^T-\underline{\underline{f}}^T-\underline{\underline{h}}^T\underline{E}_x]\underline{E}_x^{-2} = [\underline{\underline{u}}^T-\underline{\underline{f}}^T]\underline{E}_x^{-2}$$
$$\text{and } \underline{\underline{a}}^T = [\underline{\underline{u}}^T-\underline{\underline{g}}^T]\underline{E}_t^{-1}. \quad (9.25a)$$

So we have,

$$[\underline{\underline{u}}^T-\underline{\underline{g}}^T]\underline{E}_t^{-1} = \alpha^2[\underline{\underline{u}}^T-\underline{\underline{f}}^T]\underline{E}_x^{-2} - \alpha^2\underline{\underline{h}}^T\underline{E}_x^{-1}. \quad (9.25b)$$

Equating the like solumns on either side of eqn.(9.25b) we obtain the genral relationship for $u_{i,j}$ as

$$(n-2m^2\alpha^2)u_{i,j} = 8m^2\alpha^2\left[\sum_{k=1}^{i-1}(-1)^{i-k}(i-k)u_{k,j}\right] + 2m^2\alpha^2(-1)^i$$

$$\times\left[(2i-1)f_{i,j}+\frac{h_{i,j}}{2m}\right] - 2n\left[\sum_{l=1}^{j-1}(-1)^{j-l}u_{i,l}\right] - n(-1)^j g_{i,j}, \quad (9.26)$$

for all i,j with the following conditions:

(i) for i=1, the first term on the right-hand side is zero
(ii) for j=1, the third term on the right-hand side is zero
(iii) for i=1, and j=1 conditions (i) and (ii) hold jointly.

Solution for $u_{i,j}$ via eqn.(9.26) is much simpler and more convenient to implement, as we have avoided inversion of very large matrices. Care should be taken to see that $(n-2m^2\alpha^2)^{-1}$ exists. Equation (9.26) too does not give any idea about the accuracy and stability of solution.

Method (c): Recursion by single block extension

We scale the x and t axes such that $mx=\hat{x}$, and $nt=\hat{t}$ leading to

$$\frac{\partial u}{\partial \hat{t}}(\hat{x},\hat{t}) = \alpha^2\,\frac{m^2}{n}\,\frac{\partial^2 u}{\partial \hat{x}^2}(\hat{x},\hat{t}), \quad (9.27)$$

with the inital conditions along the $\hat{x}$ and $\hat{t}$ axes modified appropriately. Let $\alpha^2m^2/n=\gamma^2$. Proceeding in a similar manner as before, besides finding the piecewise constant solution for u in the region $0<\hat{x}\leq 1$ and $0<\hat{t}\leq 1$, we obtain the new initial conditions $u(1,\hat{t})$, $(\partial u/\partial x)(1,\hat{t})$ and $u(\hat{x},1)$. Then we continue the solution in either direction of $\hat{x}$ or $\hat{t}$. For this single block m=n=1 and we have

$$\underline{E}_x = \underline{E}_t = \tfrac{1}{2},\ \underline{E}_t^{-1} = \underline{E}_x^{-1} = 2 \quad\text{and}\quad \underline{E}_x^{-2} = 4\,.$$

Equations (9.24) together with (9.25) give

$$u = \frac{1}{(1-2\gamma^2)}\{-2\gamma^2 f^*_{i=o} - \gamma^2 h^*_{i=o} + g^*_{j=o}\} = u_{1,1}\,.$$

Further,

$$f^*_{i=1} = 2u - f^*_{i=o},$$
$$h^*_{i=1} = 4[u-f^*_{i=o}] - h^*_{i=o}, \qquad (9.28)$$
$$g^*_{j=1} = 2u - g^*_{j=o} \text{ (see Appendix 9.3)}.$$

Generalizing eqn.(9.28) to obtain solution for any $u_{i,j}$ over $0<x\leqq 1$ and $0<t\leqq 1$, we return to the double subscript notation and express eqn.(9.28) in a convenient matrix form as

$$\begin{bmatrix} u_{i,j} \\ f^*_{i,j} \\ h^*_{i,j} \\ g^*_{i,j} \end{bmatrix} = \begin{bmatrix} 0 & \frac{2\gamma^2}{1-2\gamma^2} & -\frac{\gamma^2}{1-2\gamma^2} & \frac{1}{1-2\gamma^2} \\ 0 & -\frac{1+2\gamma^2}{1-2\gamma^2} & -\frac{2\gamma^2}{1-2\gamma^2} & \frac{2}{1-2\gamma^2} \\ 0 & -\frac{4}{1-2\gamma^2} & \frac{1+2\gamma^2}{1-2\gamma^2} & \frac{4}{1-2\gamma^2} \\ 0 & -\frac{4\gamma^2}{1-2\gamma^2} & -\frac{2\gamma^2}{1-2\gamma^2} & \frac{1+2\gamma^2}{1-2\gamma^2} \end{bmatrix} \begin{bmatrix} 0 \\ f^*_{i-1,j} \\ h^*_{i-1,j} \\ g^*_{i,j-1} \end{bmatrix}. \qquad (9.29)$$

It is interesting to note that the eigenvalues the transition matrix are 0, +1, -1 and $-(1+2\gamma^2)/(1-2\gamma^2)$, meaning that the computation is always unstable. However for large values of γ^2, by arbitrary choice of m and n it may perhaps be possible to obtain a solution which tends to the actual one.

Table 9.2 presents the solutions obtained for j=5 by the methods (b) and (c) and compares them with the exact values of $u_{i,j}$ obtained from the exact solution of the example given below for m=25 and n=5

$$\frac{\partial u}{\partial t}(x,t) = \alpha^2 \frac{\partial^2 u}{\partial x^2}(x,t) \text{ in the region } 0\leqq x\leqq 1 \text{ and } 0\leqq t\leqq 1, \qquad (9.30)$$

with $u(0,t) = 0$, $(\partial u/\partial x)(0,t) = V_o e^{-t}$, $u(x,0) = V_o \sin x$ where $V_o = 1$ and $\alpha^2 = 1$. The exact solution is $u(x,t) = e^{-t}\sin x$.

Table 9.2. Solution of eqn.(9.30) via two-dimensional block-pulse functions. Piecewise constant values of $u_{i,5}$

i	Method (b)	Method (c)	Exact values
1	0.81414096×10^{-2}	0.81414096×10^{-2}	0.81432648×10^{-2}
2	0.24410006×10^{-1}	0.24410017×10^{-1}	0.24418868×10^{-1}
3	0.40635783×10^{-1}	0.40635839×10^{-1}	0.40653814×10^{-1}
4	0.56789692×10^{-1}	0.56789853×10^{-1}	0.56824442×10^{-1}
5	0.72841644×10^{-1}	0.72842121×10^{-1}	0.72904646×10^{-1}
6	0.88760436×10^{-1}	0.88761687×10^{-1}	0.88867128×10^{-1}
7	0.10451323	0.10451573	0.10468817
8	0.12006438	0.12006915	0.12034118
9	0.13537550	0.13538367	0.13580298
10	0.15040427	0.15041667	0.15104574
11	0.16510254	0.16512018	0.16614823
12	0.17941469	0.17943931	0.18078423
13	0.19327557	0.19330949	0.19523203
14	0.20660824	0.20665479	0.20936608
15	0.21931922	0.21938407	0.22316575
16	0.23128761	0.23138702	0.23660928
17	0.24240983	0.24252886	0.24967283
18	0.25248337	0.25264436	0.26233768
19	0.26130301	0.26152962	0.27458388
20	0.26862234	0.26893228	0.28638947
21	0.27414227	0.27453977	0.29773688
22	0.27747023	0.27796388	0.30860913
23	0.27811360	0.27872294	0.31898558
24	0.27546352	0.27621901	0.32885343
25	0.26875114	0.26971030	0.33819526

Example 9.3.

Let us consider the wave equation

$$\frac{\partial^2 u}{\partial t^2}(x,t) = \alpha^2 \frac{\partial^2 u}{\partial x^2}(x,t) \quad \text{over } 0 \leqq x \leqq 1 \text{ and } 0 \leqq t \leqq 1,$$

with (9.31)

$$u(0,t) = f(t), \ \frac{\partial u}{\partial x}(0,t) = h(t), \ u(x,0) = g(x), \ \frac{\partial u}{\partial t}(x,0) = q(x).$$

Let the BPF series expansions of $\partial^2 u/\partial t^2$ and $\partial^2 u/\partial x^2$ be

$$\frac{\partial^2 u}{\partial t^2}(x,t) = \underline{\underline{a}}^T \underline{\beta}(x,t),$$

$$\frac{\partial^2 u}{\partial x^2}(x,t) = \underline{\underline{b}}^T \underline{\beta}(x,t). \tag{9.32}$$

Also let m and n be the number of segments over the unit interval on x and t axes respectively.

Method (a): 'Kronecker product' formula

Integrating eqn.(9.30) twice with respect to 't' and twice with respect to 'x' we obtain:

$$\int_0^x \int_0^x \int_0^t \int_0^t \frac{\partial^2 u}{\partial x^2} dt^2 dx^2 = \alpha^2 \int_0^t \int_0^t \int_0^x \int_0^x \frac{\partial^2 u}{\partial x^2} dx^2 dt^2,$$

or

$$\int_0^x \int_0^x \left[u(x,t) - u(x,0) - \int_0^t \frac{\partial u}{\partial t}(x,0)\,dt\right] dx^2$$

$$= \alpha^2 \int_0^t \int_0^t \left[u(x,t) - u(0,t) - \int_0^x \frac{\partial u}{\partial x}(0,t)\,dx\right] dt^2. \tag{9.33à}$$

Rewriting eqn.(9.33) in terms of the BPF expansion and the respective operational matrices we have

$$[\underline{\underline{u}}^T - \underline{\underline{g}}^T - \underline{\underline{q}}^T \underline{E}_t] \underline{E}_x^2 = \alpha^2 [\underline{\underline{u}}^T - \underline{\underline{f}}^T - \underline{\underline{h}}^T \underline{E}_x] \underline{E}_t^2 . \tag{9.33b}$$

On rearranging, eqn. (9.33b) leads to the solution $\underline{\underline{u}}^T$ as

$$\underline{\underline{u}}^T = \{[\underline{\underline{g}}^T+\underline{\underline{q}}^T\underline{\underline{E}}_t]\underline{\underline{E}}_x^2-\alpha^2[\underline{\underline{f}}^T+\underline{\underline{h}}^T\underline{\underline{E}}_x]\underline{\underline{E}}_t^2\}(\underline{\underline{E}}_x^2-\alpha^2\underline{\underline{E}}_t^2)^{-1}. \tag{9.33c}$$

Equation (9.33c) gives the solution of $\underline{\underline{u}}^T$ provided m and n are so chosen that $(\underline{\underline{E}}_x^2-\alpha^2\underline{\underline{E}}_t^2)^{-1}$ exists. However the solution is quite cumbersome as addition, multiplication and inversion with (mnxmn) matrices are involved.

Method (b): Recurrence formula

We can express eqn. (9.30) by the use of eqn.(9.31) as

$$\underline{\underline{a}}^T\underline{\beta}(x,t) = \alpha^2\underline{b}^T\underline{\beta}(x,t) \quad \text{or } a_{i,j} = \alpha^2 b_{i,j}. \tag{9.34}$$

We have

$$\begin{aligned} \underline{\underline{a}}^T &= [\underline{\underline{u}}^T-\underline{\underline{g}}^T-\underline{\underline{q}}^T\underline{\underline{E}}_t]\underline{\underline{E}}_t^{-2} = [\underline{\underline{u}}^T-\underline{\underline{g}}^T]\underline{\underline{E}}_t^{-2}-\underline{\underline{q}}^T\underline{\underline{E}}_t^{-1}, \\ \underline{\underline{b}}^T &= [\underline{\underline{u}}^T-\underline{\underline{f}}^T]\underline{\underline{E}}_x^{-2}-\underline{\underline{h}}^T\underline{\underline{E}}_x^{-1}. \end{aligned} \tag{9.35}$$

The upper triangular structure of $\underline{E}_t^{-1}$, $\underline{E}_x^{-1}$, $\underline{E}_t^{-2}$ and $\underline{E}_x^{-2}$ (see Appendix 9.3) enables us to find a recurrence relation for $a_{i,j}$ and $b_{i,j}$ in terms of the coefficients of $\underline{\underline{u}}^T$, $\underline{\underline{g}}^T$, $\underline{\underline{q}}^T$, $\underline{\underline{f}}^T$ and $\underline{\underline{h}}^T$ respectively. This leads to a general relation for $u_{i,j}$ as

$$\begin{aligned} (n^2-\alpha^2m^2)u_{i,j} &= 4\alpha^2m^2\left[\sum_{l=1}^{i-1}(-1)^{i-1}(i-1)u_{i,j}\right] \\ &+ \alpha^2m^2(-1)^i\left[(2i-1)f_{i,j} + \frac{h_{i,j}}{2m}\right] \\ &- 4n^2\left[\sum_{k=1}^{j-k}(-1)^{j-k}(j-k)u_{i,k}\right] \\ &- n^2(-1)^j\left[(2j-1g_{i,j} + \frac{q_{i,j}}{2n}\right], \end{aligned} \tag{9.36}$$

with the following conditions:

(i) for i=1, the first term in the right-hand side is zero,

(ii) for j=1, the third term on the right-hand side is zero,

(iii) for i=1, j=1 conditions (i) and (ii) hold jointly.

The solution of $u_{i,j}$ via eqn. (9.36) is much more convenient than via eqn. (9.33c). However, m and n should be so chosen that $(n^2-\alpha^2m^2)$ does not vanish.

Method (c): Recursion by single block matrix extension

Proceeding exactly in the same manner as in the previous two examples, we change the scale of x and t such that $mx = \hat{x}$ and $nt = \hat{t}$ leading to the equation,

$$\frac{\partial^2 u}{\partial \hat{t}^2}(\hat{x},\hat{t}) = \alpha^2 \frac{m^2}{n^2} \frac{\partial^2 u}{\partial \hat{x}^2}(\hat{x},\hat{t}) = \gamma^2 \frac{\partial^2 u(\hat{x},\hat{t})}{\partial \hat{x}^2} \text{ where } \gamma^2 = \alpha^2 \frac{m^2}{n^2} \quad (9.37)$$

The initial conditions are now

$$u(0,\hat{t}),\ \frac{\partial u}{\partial \hat{x}}(0,\hat{t}),\ u(\hat{x},0) \text{ and } \frac{\partial u}{\partial \hat{t}}(\hat{x},0),$$

for the first block. Solving for the first block, we obtain in addition to $u=u_{1,1}$, the boundary conditions at $\hat{x}=1$, and $\hat{t}=1$, viz

$$u(1,\hat{t}),\ \frac{\partial u}{\partial \hat{x}}(1,\hat{t}),\ u(\hat{x},1) \text{ and } \frac{\partial u}{\partial \hat{t}}(\hat{x},1).$$

They serve as initial conditions for computation in the adjacent blocks.

The generalized equation for solution of $u_{i,j}$ corresponding to the original system by this method is

$$\begin{aligned}
u_{i,j} &= -\frac{\gamma^2}{1-\gamma^2}[f^*_{i-1,j} + \tfrac{1}{2}h^*_{i-1,j}] + \frac{1}{1-\gamma^2}[g^*_{i,j-1} + \tfrac{1}{2}q^*_{i,j-1}],\\
f^*_{i,j} &= 2u_{i,j} - f^*_{i-1,j},\\
h^*_{i,j} &= 4[u_{i,j} - f^*_{i-1,j}] - h^*_{i-1,j}, \qquad (9.38)\\
g^*_{i,j} &= 2u_{i,j} - g^*_{i,j-1},\\
q^*_{i,j} &= 4[u_{i,j} - g^*_{i,j-1}] - q^*_{i,j-1},
\end{aligned}$$

Equations (9.38) can be written in a matrix form as:

$$\begin{bmatrix} u_{i,j} \\ f^*_{i,j} \\ h^*_{i,j} \\ g^*_{i,j} \\ q^*_{i,j} \end{bmatrix} = \begin{bmatrix} 0 & -\frac{\gamma^2}{1-\gamma^2} & -\frac{\gamma^2}{2(1-\gamma^2)} & \frac{1}{1-\gamma^2} & \frac{1}{2(1-\gamma^2)} \\ 0 & -\frac{1+\gamma^2}{1-\gamma^2} & -\frac{\gamma^2}{1-\gamma^2} & \frac{2}{1-\gamma^2} & \frac{1}{1-\gamma^2} \\ 0 & -\frac{4}{1-\gamma^2} & -\frac{1+\gamma^2}{1-\gamma^2} & \frac{4}{1-\gamma^2} & \frac{2}{1-\gamma^2} \\ 0 & -\frac{2\gamma^2}{1-\gamma^2} & -\frac{\gamma^2}{1-\gamma^2} & \frac{1+\gamma^2}{1-\gamma^2} & \frac{1}{1-\gamma^2} \\ 0 & -\frac{4\gamma^2}{1-\gamma^2} & -\frac{2\gamma^2}{1-\gamma^2} & \frac{4\gamma^2}{1-\gamma^2} & \frac{1+\gamma^2}{1-\gamma^2} \end{bmatrix} \begin{bmatrix} 0 \\ f^*_{i-1,j} \\ h^*_{i-1,j} \\ g^*_{i,j-1} \\ q^*_{i,j-1} \end{bmatrix} \qquad (9.39)$$

The solution of $u_{i,j}$ via eqn. (9.39) is the most convenient of all. Further it shows us that the solution obtained is not computationally stable since the transition matrix has eigenvalues of +1, +1, 0, -1, -1. The double roots at +1 and -1 are the causes of instability. The eigenvalues are not dependent on m and n. The solution via eqn. (9.39) is the same as that obtained by the eqn. (9.33) and (9.36).

As an illustration let us consider a flexible string of length π, tightly stretched between points $l=0$ and $l=\pi$ along the axis, its ends fixed at these points. When set into small transverse vibrations the displacement $u(l,t)$ from the initial position at any time t and at any point l is given by

$$\frac{\partial^2 u}{\partial t^2} = \alpha^2 \frac{\partial^2 u}{\partial l^2} , \qquad (9.40)$$

where

$$\alpha^2 = 4, \ u(0,t) = 0 ,$$

$$\frac{\partial u}{\partial t}(0,t) = 0.1 \cos 2t + 0.04 \cos 8t ,$$

$$u(l,0) = 0.1 \sin l + 0.01 \sin 4l ,$$

$$\frac{\partial u}{\partial t}(l,0) = 0 \quad \text{for } 0 \leq l \leq \pi , \ t>0 .$$

The exact solution is

$$u(l,t) = 0.1 \sin l \cos 2t + 0.01 \sin 4l \cos 8t \quad . \qquad (9.41)$$

Table 9.3: Solution of eqn. (9.40) via two-dimensional block-pulse functions. Piecewise constant values of $u_{i,1}$

i	Method (b)	Method (c)	Exact values
1	$0.10862368\text{x}10^{-1}$	$0.10862380\text{x}10^{-1}$	$0.10783892\text{x}10^{-1}$
2	$0.30411378\text{x}10^{-1}$	$0.30441378\text{x}10^{-1}$	$0.31028826\text{x}10^{-1}$
3	$0.49713902\text{x}10^{-1}$	$0.49713802\text{x}10^{-1}$	$0.47741283\text{x}10^{-1}$
4	$0.50026875\text{x}10^{-1}$	$0.50027087\text{x}10^{-1}$	$0.59856247\text{x}10^{-1}$
5	0.11118007	0.11117893	$0.67728996\text{x}10^{-1}$
6	-0.12328541	-0.12328041	$0.72876453\text{x}10^{-1}$
7	0.96073359	0.96071064	$0.77286839\text{x}10^{-1}$
8	$-0.38961039\text{x}10^{1}$	$-0.38960009\text{x}10^{1}$	$0.82562983\text{x}10^{-1}$
9	$0.18008957\text{x}10^{2}$	$0.18008514\text{x}10^{2}$	$0.89227200\text{x}10^{-1}$
10	$-0.80611710\text{x}10^{2}$	$-0.80609680\text{x}10^{2}$	$0.96463919\text{x}10^{-1}$
11	$0.36360132\text{x}10^{3}$	$0.36359229\text{x}10^{3}$	0.10238224
12	$-0.16370513\text{x}10^{4}$	$-0.16370112\text{x}10^{4}$	0.10472226
13	$0.73736445\text{x}10^{4}$	$0.73734609\text{x}10^{4}$	0.10171473
14	$-0.33209453\text{x}10^{5}$	$-0.33208629\text{x}10^{5}$	$0.92781842\text{x}10^{-1}$
15	$0.14957169\text{x}10^{6}$	$0.14956806\text{x}10^{6}$	$0.78795254\text{x}10^{-1}$
16	$-0.67365219\text{x}10^{6}$	$-0.67363563\text{x}10^{6}$	$0.61811086\text{x}10^{-1}$
17	$0.30340450\text{x}10^{7}$	$0.30339710\text{x}10^{7}$	$0.4436193\text{x}10^{-1}$
18	$-0.13664971\text{x}10^{8}$	$-0.13664633\text{x}10^{8}$	$0.28589554\text{x}10^{-1}$
19	$0.61545360\text{x}10^{8}$	$0.61543808\text{x}10^{8}$	$0.15534699\text{x}10^{-1}$
20	$-0.27719270\text{x}10^{9}$	$-0.27718554\text{x}10^{9}$	$0.48657991\text{x}10^{-2}$

Let us change the scale of "l" such that $l=\pi x$. This transforms the given equation to

$$\frac{\partial^2 u}{\partial t^2} = \frac{4}{\pi^2}\frac{\partial^2 u}{\partial x^2} \text{, for } t>0 \quad \text{and } 0 \leqq x \leqq 1 \; . \qquad (9.42)$$

The new initial conditions are

$$u(x,0) = 0.1 \sin\pi x + 0.01 \sin 4\pi x, \quad \frac{\partial u}{\partial t}(x,0) = 0 ,$$

and the boundary conditions at x=0 are

$$u(0,t) = 0, \frac{\partial u}{\partial x}(0,t) = 0.1\pi\cos 2t + 0.4\pi \cos 8t .$$

Solution of eqn. (9.42) is obtained for $0 \leq x \leq 1$ and $0 \leq t \leq 1$ with m=20 and n=20. Table 9.3 gives the solution of $u_{i,j}$ for j=1, by the methods (b), and (c) and compares it with the values obtained from the exact solution (eqn. (9.41)).

9.2. Conclusion

A method of numerically integrating partial differential equations is presented. The technique employs multidimensional block pulse functions and yields results with minimal mean square error. The general solutions are reduced to simple recursive form. Solutions in such simple recursive form are due to the disjoint nature of block pulse functions. An important feature of the recursive solutions via BPF's is that an insight into the stability of the computation is gained. Three types of equation have been considered. The first order partial differntial equation is computationally stable by the present method. The two second-order equations do not promise absolute stability. Palanisamy [W36] inserted the OSOMRI of single dimension (of section 2.3.5) in the Kronecker products leading to multidimensional OSOMRI and used these in solving PDE of second order. Extensive computations have been made with Examples 9.2 and 9.3 with the multidimensional OSOMRI. Although an improvement in the accuracy has been reported [W36], the inherent instability of computations in the two second order examples remained without improvement. In conclusion, it is reasonable to state that a careful stability analysis should precede the use of piecewise constant orthogonal functions in the solutions of partial differential equations as natural extensions of the previous results in every case.

9.3. Appendix

Relationship between one-dimensional (sections along the grid line) and two dimensional BPF coefficients of functions of two variables.

Let us consider the function $u(\hat{x},\hat{t})$ and its partial derivative $(\partial u/\partial \hat{x})(\hat{x},\hat{t})$ expressed in BPF expansion as

$$u(\hat{x},\hat{t}) = \underline{\underline{u}}^T \underline{\beta}(\hat{x},\hat{t}) = [u_{1,1} \ldots u_{1,n} \ldots u_{m,1} \ldots u_{m,n}] \underline{\beta}(\hat{x},\hat{t}) \quad (9.A1)$$

$$\frac{\partial u}{\partial \hat{x}}(\hat{x},\hat{t}) = \underline{\underline{a}}^T \underline{\beta}(\hat{x},\hat{t}) = [a_{1,1} \ldots a_{1,n} \ldots a_{m,1} \ldots a_{m,n}] \underline{\beta}(\hat{x},\hat{t}) \quad (9.A2)$$

Let

$$\left.\begin{aligned} f^*_{i,j} &= \text{BPF value of } u(\hat{x},\hat{t}), && \text{at } \hat{x}=i,\ \hat{t}\in(j-1,j] \\ g^*_{i,j} &= \text{BPF value of } u(\hat{x},\hat{t}), && \text{at } \hat{t}=j,\ \hat{x}\in(i-1,i] \\ h^*_{i,j} &= \text{BPF value of } \frac{\partial u}{\partial \hat{x}}(\hat{x},\hat{t}), && \text{at } \hat{x}=i,\ \hat{t}\in(j-1,j] \\ q^*_{i,j} &= \text{BPF value of } \frac{\partial u}{\partial \hat{t}}(\hat{x},\hat{t}), && \text{at } \hat{t}=j,\ \hat{x}\in(i-1,j] \end{aligned}\right\} \quad (9.A3)$$

for all $i=0,1,2,\ldots,m$ and $j=0,1,2,\ldots,n$.

In the interval $0<\hat{t}\leq 1$, the one dimensional BPF value of $u(\hat{x},\hat{t})$ is a function of $\hat{x}$. So we have

$$\text{BPF value of } \int_0^{\hat{x}} \frac{\partial u}{\partial \hat{x}}(\hat{x},\hat{t})d\hat{x} = \text{BPF value of } u(\hat{x},\hat{t}) \quad (9.A4)$$

$$- \text{BPF value of } u(0,\hat{t}) = f^*_{1,1} - f^*_{0,1}, \quad \text{for } t\in(0,1].$$

The use of (9.A2) and the replacement of integration by summation on the left-hand side of (9.A4) for m=1 yields

$$a_{1,1} = f^*_{1,1} - f^*_{0,1}. \quad (9.A5)$$

The use of (9.A1) and (9.A2) in the identity

$$\int_0^x \frac{\partial u}{\partial \hat{x}}(\hat{x},\hat{t})d\hat{x} = u(\hat{x},\hat{t}) - u(0,\hat{t}),$$

yields for $\hat{t} \in (0,1]$,

$$\frac{1}{2} a_{1,1} = u_{1,1} - \text{two dimensional BPF value of } u(0,\hat{t})$$

$$= u_{1,1} - f_{1,1} = u_{1,1} - f^*_{0,1} \qquad (9.A6)$$

Eliminating $a_{1,1}$ from (9.A5) and (9.A6),

$$f^*_{1,1} = 2u_{1,1} - f^*_{0,1} \, . \qquad (9.A7)$$

Similarly, we can show that

$$g^*_{1,1} = 2u_{1,1} - g^*_{1,0} \, , \qquad (9.A8)$$

$$h^*_{1,1} = 4[u_{1,1} - f^*_{0,1}] - h^*_{0,1} \, , \qquad (9.A9)$$

$$q^*_{1,1} = 4[u_{1,1} - g^*_{1,0}] - q^*_{1,0} \, . \qquad (9.A10)$$

In general, it can be shown that

$$\left.\begin{aligned} f^*_{i,j} &= 2u_{i,j} - f^*_{i-1,j} \, , \\ g^*_{i,j} &= 2u_{i,j} - g^*_{i,j-1} \, , \\ h^*_{i,j} &= 4[u_{i,j} - f^*_{i-1,j}] - h^*_{i-1,j} \, , \\ q^*_{i,j} &= 4[u_{i,j} - g^*_{i,j-1}] - q^*_{i,j-1} \, . \end{aligned}\right\} \qquad (9.A11)$$

Equations (9.A11) are employed in method (c).

X

IDENTIFICATION OF CONTINUOUS LUMPED PARAMETER SYSTEMS

10.1. Introduction

Parameter identification in continuous lumped parameter systems is of considerable interest to control engineers. Most of the actual physical models encountered in practice are basically continuous. In view of the operating mode of almost all digital process control computers, system models are deliberately discretized. Although the process of discretization is through a certain approximation, the convenience with which the discrete models lend themselves to various methods of analysis, identification, design and implementation of control algorithms outweighs all the undesirable effects of approximation in discretization. There is yet another important reason for resorting to discretization of models. The derivative measurment problem in the identification of continuous models has been a major deterrent. This problem is elegantly overcome in the methods based on PCBF characterized process data. Walsh functions as the basis for process data characterization have dominated the scene in the recent years. Prasada Rao and Sivakumar [W 46-W 50], Chen and Hsiao [W 13], Tzafestas [W 70], Karanam, Frick and Mohler [W 30] and Prasada Rao and Palanisamy [W 44] have made important contributions to the field of identification of continuous lumped systems. In this chapter, we will discuss the main technique and its several variants applicable to different situations. Our main vehicle in the course of the present treatment will be Walsh functions. Other basis functions, particularly the BPF have also been used [W 38, W 28]. There is no difference in the philosophy, and the format of the algorithm is the same as with WF. Some computational advantages have been claimed [W 28].

The problem of parameter estimation in continuous systems from continuous input-output data is also tackled by a hybrid approach in two steps. At first a discrete model is estimated from sampled data. This discrete model is then converted into continuous form. The conversion of the discrete model into continuous (Laplace transform) form is not without difficulties. The difficulties in this sampled data approach [I 11, I 12, I 21, I 22] are mainly due to loss of inter-sample information. Attemps to remedy the situation rely on a priori knowledge of the information between sampling instants. At the beginning of the procedure, sampling has to be arbitrarily fast if one does not know initially the approximate range of values of the various modes of the system under identification. Some guidelines may be drawn from the theory of digital filters for conversion of the discrete model into continuous form. Sinha [I 21] suggests bilinear transformation as a method suitable for a large number of practical situations for conversion. There are no attempts, as for as the author is aware , to campare and contrast the two approaches viz., the two-stage technique via discrete model with sampled data and the direct continuous-data-continuous-model method.

In recent years, the WF method to avoid noise accentuating derivative operations directly on process data in estimating continuous models, has been highly successful. The WF characterization of signals reduces the calculus of continuous systems to an elegant, optimally (in the sense of least squares) approximate algebra through the operational matrices discussed in chapter II. Models which are linear in their parameters give rise to linear algebraic equations in terms of system parameters and unknown initial conditions. Walsh spectral analysers [W 24] may be physically connected to the system under identification for practical on-line implementation. Apart from avoiding direct derivative operations on process data, the WF method is computationally attractive and inherently resistive to zero-mean additive noise [W 49]. Hardware for WF processing is simple and rugged as it involves standard switching elements.

The WF technique was first introduced by Prasada Rao and Sivakumar [W 46] and Chen and Hsiao [W 13] in system identification problems for lumped models. Tzafestas [W 70], Karanam, Frick and Mohler [W 30] and Chen and Shih [W 19] gave algorithms for other kinds of lumped models. Tzafestas [W 70], Paraskevoloulos and Bounas [W 39] and Sinha et. al [W 64] apply the basic ideas of the WF technique extended to distributed parameter systems. Prasada Rao and Sivakumar considered problems with unknown time delays

[W 47], multiple-input-multiple-output (mimo) models [W 48], piecewise linear models [W 50] and problems of order identification [W 49].

Recently Prasada Rao and Palanisamy [W 44] improved all the existing WF algorithms in terms of accuracy employing the OSOMRI. Using Walsh delay matrices the algorithm for time-lag models is renderend direct and non-iterative when the time-delays are known to be small. When the delay is not small, the results of the direct algorithm [W 44] serve as good initiatives for the inevitable iterative shift algorithm [W 47]. At first we discuss the basic algorithm for parameter identification in lumped linear time-invariant continuous models.

10.2. Basic algorithm for parameter identification in lumped linear time invariant systems via WF (BAPILLTIS)

Consider a system modelled by:

$$y(t) + \sum_{i=1}^{n} a_i \frac{d^i y(t)}{dt^i} = b_o\, x(t) + \sum_{j=1}^{n-1} b_j \frac{d^i x(t)}{dt^i} , \qquad (10.1)$$

in which the input $x(t)$, and the output $y(t)$ are available over an arbitrary but active period of time. The term 'active' is used to mean that over the record the input is able to excite all the modes of the system. The essential parameters to be estimated are: $[a_i,\ i = 1,2,\dots,\ n]$ and $[b_j,\ j = 0,1,2,\dots,\ n-1]$. Owing to the arbitrariness of the record, there are also other unkowns:

$$[\alpha_{fi} \triangleq \left.\frac{d^{i-1}y}{dt^{i-1}}\right|_{t=0} \quad i = 2,3,\dots,\ n] \text{ and } [\beta_{xi} \triangleq \left.\frac{d^{i-1}x}{dt^{i-1}}\right|_{t=0} \quad i = 2,3,..n-1].$$

Although $x(0)$ and $y(0)$ are actually known, we will ignore this information and treat these too as unknowns along with the other initial conditions.

The length of the record may be normalized to form a unit interval by suitable time scaling. Let the data record length be normal (or normalized). In the normalized situation, integrate equation (10.1) n times and expand the various terms in Walsh series of size m. Then we equate the like sequency components on either side and get a set of simultaneous equations with the process parameters and initial conditions as the unknowns. The equations may be arranged in the general form:

$$\underline{\Phi}_{yxt}\ \underline{u}_{abf} = \underline{c}_y \quad , \tag{10.2}$$

where

$$\underline{\Phi}_{yxt} = [\underline{\Phi}_y \mid \underline{\Phi}_x \mid \underline{\mathcal{J}}^T] \quad , \tag{10.3a}$$

$$\underline{u}_{abf} = [a_1,\ a_2,\ \dots\ a_n,\ b_o,\ b_1,\ \dots\ b_{n-1},\ f_1,\ f_2,\ \dots\ f_n]$$

$$\triangleq [\underline{a}^T \mid \underline{b}^T \mid \underline{f}^T] \quad , \tag{10.3b}$$

$$f_i = \underline{\alpha}_y^T\ \underline{\Delta}^{n-i}\ \underline{a} + \underline{\beta}_x^T\ \underline{\Delta}^{n-i-1}\ \underline{b} \quad , \tag{10.3c}$$

$$\underline{\alpha}_y = -[\alpha_{y1},\ \alpha_{y2},\ \dots\ \alpha_{yn}]^T \quad , \tag{10.3d}$$

$$\underline{\beta}_x = [\beta_{x1},\ \beta_{x2},\ \dots\ \beta_{xn}]^T \quad , \tag{10.3e}$$

$$\underline{\Delta} = \begin{bmatrix} 0 & 1 & 0 & 0 & \dots & 0 \\ 0 & 0 & 1 & 0 & \dots & 0 \\ & \dots & & & \dots & \\ 0 & 0 & 0 & 0 & \dots & 1 \\ 0 & 0 & 0 & 0 & \dots & 0 \end{bmatrix} \quad , \tag{10.3f}$$

$$\underline{\Phi}_y = [\underline{E}_{n-1}^T\ \underline{y} \mid \underline{E}_{n-2}^T\ \underline{y} \mid \dots \mid \underline{y}] \quad , \tag{10.3g}$$

$$\underline{\Phi}_x = -[\underline{E}_n^T\ \underline{x} \mid \underline{E}_{n-1}^T\ \underline{x} \mid \dots \mid \underline{E}^T\ \underline{x}] \quad , \tag{10.3h}$$

$$\underline{c}_y = -\underline{E}_n^T\ \underline{y} \quad , \tag{10.3i}$$

$$y(t) = \underline{y}^T\ \underline{w}(t) \quad , \tag{10.3j}$$

$$x(t) = \underline{x}^T\ \underline{w}(t) \quad , \tag{10.3k}$$

$$\underline{y} = [y_o,\ y_1,\ \dots\ y_{m-1}]^T \quad , \tag{10.3l}$$

$$\underline{x} = [x_o,\ x_1,\ \dots\ x_{m-1}]^T \quad , \tag{10.3m}$$

$$y_i = \int_o^1 y(t)\ w_i(t)\ dt \quad , \tag{10.3n}$$

$$x_i = \int_o^1 x(t)\ w_i(t)\ dt, \text{ and} \tag{10.3p}$$

$$\underline{E}_k = \underline{E}^k \quad . \tag{10.3q}$$

We drop the subscript w from $\underline{E}_w$ for the sake of simplicity since our treatment is understood to be wholly with WF.

$\underline{\mathcal{J}}$ has elements which happen to be Walsh components of terms of the form t^i/i. In fact,

$$\underline{\mathcal{J}}_{(n\times m)} = \begin{bmatrix} \text{I st row of } \underline{E}^0 \\ \hline \text{I st row of } \underline{E} \\ \hline \ldots \quad \ldots \\ \hline \text{I st row of } \underline{E}^{n-1} \end{bmatrix} \quad . \tag{10.3r}$$

Equation (10.2) may be now solved for $\underline{u}_{abf}$. When the amount of information is more than necessary, we may get $\underline{u}_{abf}$ by least squares method. This algorithm will be referred to as BAPILLTIS in latter sections.

Example 10.1

In a system described by

$$\frac{d^2y}{dt^2} + a_1\frac{dy}{dt} + a_2\,y = x(t) \quad ,$$

with $a_1 = 3$, $a_2 = 2$, $x(t) = t$ and all zero initial conditions, the output $y(t)$ is generated. The portion of the input-output record from t = 1 to 2 sec. is considered. At the instant corresponding to the beginning of the chosen record the system state is taken as unknown. The identification equations are normalised to get

$$\begin{bmatrix} 0.011053644 & 0.003568715 & 0.08448329 & 0.059899638 \\ 0.003568713 & 0.0011863039 & 0.02458365 & 0.018838278 \\ 0.08448329 & 0.02458365 & 1.00000000 & 0.50000000 \\ 0.059899638 & 0.018838278 & 0.50000000 & 0.33203125 \end{bmatrix}$$

$$\begin{bmatrix} a_1 \\ a_2 \\ f_o \\ f_1 \end{bmatrix} = \begin{bmatrix} 0.0061606824 \\ 0.0025088628 \\ 0.007035840 \\ 0.025491285 \end{bmatrix} .$$

In this case $f_o = x(0)$ and $f_1 = a_1\ x(0) + \dot{x}(0)$. The following results are abtained from the above

$$a_1 = 3.009 \ , \qquad a_2 = 1.985512 \ ,$$

$$x(0) = 0.083791346, \quad \dot{x}(0) = 0.2004042 \ .$$

10.3. Transfer Function Matrix Identification in MIMO Systems

There is very little published work on the problem of determining the transfer function matrix (TFM) in linear, time-invariant continuous multi-input multi-output (MIMO) systems from a finite, arbitrary but active record of input output data. Recently Mathew and Fairman [P8] gave a method for TFM identification based on Diamessis's multiple integration technique [P1, P2]. This section presents an identification algorithm employing Walsh functions as a successful and systematic extension of the earlier works of the author with Sivakumar [W 46, W 47, W 49] to the case of MIMO systems.

Consider a continuous, linear, time-invariant MIMO system characterized by

$$\sum_{i=0}^{n_k} a_{ik} \frac{d^i y_k(t)}{dt^i} = \sum_{l=1}^{R} \sum_{j=0}^{n_k-1} b_{jl}^k \frac{d^j x_l(t)}{dt^i}, \quad k = 1,2,\dots M, \tag{10.4}$$

where $y_k(t)$ and $x_l(t)$ are the k-th output and l-th input respectively. Laplace transform of (10.4) gives:

$$\underline{\underline{Y}}(s) = \underline{\underline{F}}(s)\ \underline{\underline{X}}(s) \quad , \tag{10.5}$$

where

$$\underline{\underline{X}}(s) = [X_1(s), X_2(s) \quad \dots \quad X_R(s)]^T \quad ,$$

$$\underline{\underline{Y}}(s) = [Y_1(s), Y_2(s) \quad \dots \quad Y_M(s)]^T \quad ,$$

$$\underline{\underline{F}}(s) = \begin{bmatrix} \frac{z_{11}(s)}{D_1(s)} & \frac{z_{12}(s)}{D_1(s)} & \dots & \frac{z_{1R}(s)}{D_1(s)} \\ \frac{z_{21}(s)}{D_2(s)} & \frac{z_{22}(s)}{D_2(s)} & \dots & \frac{z_{2R}(s)}{D_2(s)} \\ \dots & \dots & & \dots \\ \frac{z_{M1}(s)}{D_M(s)} & \frac{z_{M2}(s)}{D_M(s)} & \dots & \frac{z_{MR}(s)}{D_M(s)} \end{bmatrix} \quad , \tag{10.6}$$

$$D_k(s) = \sum_{i=o}^{n_k} a_{ik}\, s^i \quad , \qquad a_{n_k k} = 1.0 \quad ,$$

and

$$z_{kl}(s) = \sum_{j=o}^{n_k-1} b_{jl}^k\, s^i \quad .$$

System identification requires the determination of $\{a_{ik}, b_{jl}^k\}$, $i = 0,1, \dots (n_k-1)$; $k = 1,2, \dots M$; $j = 0,1,\dots n_k-1$; $l = 1,2,\dots R$, using arbitrary (but active) lengths of records of input-output data. In this situation the initial conditions form a set of additional unknowns to be determined simultaneously with the essential system parameters. Although the initial conditions $y_i(0)$, $x_k(0)$ are actually known, we would deliberately ignore this information and include these in the set of unknowns for reasons given in [W 47].

Integrating (10.4) L times, L being the upper bound on the elements $\{n_k\}$, and inserting m-size Walsh spectral expansions of the various terms, we get a set of simultaneous equations linear in the unknowns. These equations may be arranged in the form:

$$\underline{\Phi}_{yxt_k}\, \underline{u}_{abf_k} = \underline{c}_{y_k} \quad , \quad k = 1,2,\dots M \tag{10.7}$$

where

$$\underline{\Phi}_{yxt_k} = [\underline{\Phi}_{y_k} \mid \underline{\Phi}_x \mid \underline{\mathcal{J}}^T] \quad , \tag{10.8a}$$

$$\underline{\Phi}_x = [\underline{\Phi}_{x_1} \mid \underline{\Phi}_{x_2} \mid \dots \mid \underline{\Phi}_{x_R}] \quad , \tag{10.8b}$$

$$\underline{\Phi}_{y_k} = [\underline{E}_{L-1}^T \underline{y}_k \mid \underline{E}_{L-2}^T \underline{y}_k \mid \dots \mid \underline{E}_{L-n_k}^T \underline{y}_k] \, , \tag{10.8c}$$

$$\underline{\Phi}_{x_j} = [\underline{E}_L^T\, \underline{x}_j \mid \underline{E}_{L-1}^T \underline{x}_j \mid \dots \mid \underline{E}_{L-n_k+1}^T \underline{x}_j] \, , \tag{10.8d}$$

$$\underline{c}_k = -\underline{E}_L^T\, \underline{y}_k \, , \tag{10.8e}$$

$$\begin{aligned} \underline{u}_{abf_k} = [&a_{o,k}, a_{1,k},\dots,a_{n_k-1,k}, -b_{0,1}^k, -b_{1,1}^k,\dots,-b_{n_k-1,1}^k, \\ &-b_{0,2}^k, -b_{1,2}^k,\dots,-b_{n_k-1,2}^k,\dots, \\ &-b_{0,R}^k, -b_{1,R}^k,\dots,-b_{n_k-1,R}^k, f_{k,1}, f_{k,2},\dots,f_{k,n_k}]^T, \end{aligned} \tag{10.8f}$$

$$\triangleq [\underline{a}_k^T \mid \underline{b}_k^T \mid \underline{f}_k^T]^T, \tag{10.8g}$$

$$\underline{b}_k = [\ldots | \underline{b}_{k,j}^T | \ldots]^T , \tag{10.8h}$$

$$\underline{b}_{k,j} = [-b_{0,j}, -b_{1,j}^k, \ldots, -b_{n_k-1,j}^k]^T , \tag{10.8i}$$

$$f_{k,j} = \underline{\alpha}_{yk}^T \underline{\Delta}^{n_k-j} \underline{a}_k + \sum_{i=1}^{R} \underline{\beta}_{xi}^T \underline{\Delta}^{n_k-j-1} \underline{b}_{k,i}^T , \tag{10.8j}$$

$$\underline{\alpha}_{yk} = -[\alpha_{yk,1}, \alpha_{yk,2}, \ldots, \alpha_{yk,n_k}]^T , \tag{10.8k}$$

$$\underline{\beta}_{xi} = [\beta_{xi,1}, \beta_{xi,2}, \ldots, \beta_{xi,n_k-1}]^T , \tag{10.8l}$$

$$\underline{\alpha}_{yk,j} = \left. \frac{d^{j-1}y(t)}{dt^{j-1}} \right|_{t=0} , \tag{10.8m}$$

$$\underline{\beta}_{xi,j} = \left. \frac{d^{j-1}x_i(t)}{dt^{j-1}} \right|_{t=0} , \tag{10.8n}$$

and

$$\underline{E}_k = \underline{E}^k . \tag{10.8p}$$

$\underline{Y}_k$ and $\underline{x}_j$ are m-vectors of Walsh spectral components of $y_k(t)$ and $x_j(t)$ respectively.

The subscript 'w' has been dropped throughout in the above to limit the complexity of notation, since our treatment should be understood to be wholly with WF alone.

The matrix $\underline{\mathcal{J}}$ in this situation is given by

$$\underline{\mathcal{J}}_{(L\times m)} = \begin{bmatrix} \text{1st row of } \underline{E}^0 \\ \text{-------------} \\ \\ \text{-------------} \\ \text{1st row of } \underline{E}^{L-1} \end{bmatrix} \tag{10.9}$$

Solution of (10.7) gives the system parameters. Notice that Φ_{yxt_k} is invertible provided the Walsh spectra of $x_j(t)$ and the rows of $\underline{\mathcal{J}}$ are linealy independent. This requirement is in the spirit of L-suitability as discussed by Mathew and Fairman [P 8]. If one considers Walsh spectral vectors larger than required, least squares technique would reduce the set of equations (10.7) into normal form.

Example 10.2: Consider the 2-input 1-output model.

$$\ddot{y}(t) + a_1\dot{y}(t) + a_o y(t) = b_{1,1}\dot{x}_1(t) + b_{0,1}x_1(t) + b_{1,2}\dot{x}_2(t) + b_{0,2}x_2(t).$$

The corresponding TFM is given by

$$\underline{\underline{F}}(s) = \left[\frac{b_{1,1}s + b_{0,1}}{s^2+a_1 s+a_o} \; , \; \frac{b_{1,2}s + b_{0,2}}{s^2+a_1 s+a_o}\right] .$$

With $b_{1,1} = 1.0$, $b_{0,1} = 0.5$, $b_{1,2} = 0.0$, $b_{0,2} = 1.0$, $a_1 = 1.5$, $a_o = 0.5$ and zero initial conditions input-output data is generated over (0.0, 1.0 sec) for a pair of L-suitable input signals. Choosing $m = 8$ the 8-vectors of Walsh spectral components so obtained are inserted in (10.7). Solving this equation the system TFM is found to be

$$\underline{\underline{F}}(s) = \left[\frac{0.999s + 0.5}{s^2+1.5s+0.5} \; , \; \frac{0.0001s + 1}{s^2+1.5s+0.5}\right] .$$

Example 10.3: Let us consider the model in Example 10.2 with an arbitrary record length of data on (0.25, 1.0 sec.). This case is of practical significance and involves the unknown initial conditions in the identification algorithm. Now the total number of unknowns is 8. Parameter identification has been carried out considering 8, 16, 32 size Walsh spectral vectors. The results are summarized in Table 10.1.

TFM identification in MIMO Systems via Walsh functions in the presence of unknown initial conditions	
No. of terms in each Walsh series of the algorithm	Obtained $\underline{F}(s)$
8	$\frac{s + 0.509}{s^2+1.509s+0.501}$, $\frac{0.0001s+1.003}{s^2+1.509s+0.501}$
16	$\frac{0.9999s+0.4994}{s^2+1.499s+0.4996}$, $\frac{0.0001s+0.9997}{s^2+1.499s+0.4996}$
32	$\frac{0.9999s+0.499}{s^2+1.499s+0.4996}$, $\frac{0.0001s+0.9997}{s^2+1.499s+0.4996}$

Table 10.1.

Notice that even the minimum number of Walsh components have been able to yield very good results hardly improved by considering additional high sequency Walsh series terms.

10.4. Order and parameter identification in continuous linear systems via Walsh functions [I26, I5, P20, I27]

System modelling from input-output data is generally carried out in two steps. The first is to determine the form of the model. In the second step the parameters of the model in the appropriate form are estimated by fitting the process data. The model may be discrete or continuous. In the existing literature many techniques assume the model order and discuss only parameter identification. The problem of order determination has been discussed by Young et al [I26] and Desai and Fairman [I5]. Desai and Fairman employ the Poisson moment functional technique in which the decision on model order is based on (i) minimizing a measure of consistency of the parameters for low noise levels, and (ii) the instrumental variable approach for higher levels of noise. This approach is general and does not require special modifications to account for the presence of zeros and multiple poles. Apparently it emphasizes the aspect of order determination alone as no parameter values are presented in any case.

In the proposed algorithm we follow the steps given below:

Step 1: From the data: $x(t)$, $y(t)$ available on $[0,T_o)$, consider subintervals I_i, viz. $[0,T_i)$, $i=1,2,\ldots,L$, $T_i \in [0,T_o)$. Obtain Walsh spectral expansions on each I_i. Normalization of I_i will be convenient. This can be done by suitable time scaling.

Step 2: Form equations (10.2) through relations detailed in (10.3a-q) for a chosen n for all i.

Step 3: Solve (10.2) for $\underline{u}_{abf_i}$, the set of parameters so obtained from data cver I_i.

Step 4: Compute the error function

$$H_n(n) = \sum_{i=1}^{L-1} \|\underline{u}_{abf_{i+1}} - \underline{u}_{abf_i}\| \qquad (10.10)$$

Step 5: Minimise H_n with respect to n and get n^o, the most appropriate system size.

The above method is applied in the following case.

Example 10.4.

A process described by

$$a_2 \frac{d^2y(t)}{dt^2} + a_i \frac{dy(t)}{dt} + y(t) = b_o x(t) , \qquad (10.11)$$

has been simulated with $a_2 = 1$, $a_1 = 1$, $b_o = 1$, zero initial conditions, and $x(t)$ as shown in Fig. 10.1. 32 Walsh components have been taken on each I_i, $i=1,2,3$ (i.e. $L=3$). It was found by direct search that $n^o = 2$. The minimum value of the error function

$$H_n(n^o) = 0.667 \times 10^{-6}, \quad n^o = 2.$$

There is a sharp drop in H_n from $n=1$, to $n=2$. In fact, $H_n(1)=0.1045$. Attempts to fit oversized models lead to numerical difficulties particularly while inverting the matrix in (10.2). This phenomenon of weakening determinant may also aid our decision process in order determination. The parameters corresponding to $n^o=2$ are: $a_2=0.9987$, $a_1=0.9983$, $b_o=0.9990$, $b_1=0.00035$, $b_2=-0.00067$. An interesting feature of the Walsh function technique noticed is that the results are quite good even with 8 Walsh components and could hardly be improved by increasing the size of the Walsh spectrum.

10.5. Iterative shift algorithm via Walsh functions for time-lag systems

A SISO linear time-invariant time lag system may, in general, have multiple delays. For the sake of a simple illustrative development, however, we consider the following particular case of a model with a single delay:

$$a_o y(t) + \sum_{i=1}^{n} a_i \frac{d^{n-i}y(t)}{dt^{n-i}} = b\, x(t-\tau), \qquad (10.12)$$

in which a_i, and τ are unknown, to be determined from $y(t)$ and $x(t)$ available in $[0,T_o)$.
The initial state of the system except $y(\cdot)$ is also unknown along with a_i and τ. Although $y(\cdot)$ is known in $[0,T_o)$ we will treat it as unknown.

The identification algorithms may be applied in the following steps:

1. Start with a guess value of τ and denote $x(t-\tau)$ as $x_d(t)$.

2. Consider subintervals $I_i \triangleq [t_s, t_i)$, $i=1,2,\ldots,L$. $t_i \in [t_s, T_o)$, $t_s \geq \tau$, after the beginning of the available data record. Notice that τ is yet an unknown. We should choose t_s sufficiently inside the interval $[0, T_o)$ based on the maximum expected value for τ. Beginning all I_i at $t=t_s$ is to retain the same initial conditions.

3. Choose m-sets of WF on each I_i and determine the Walsh spectra of $y(t)$ and $x_d(t)$ in each interval. In each case, the width of the base of WF should be accounted for in Walsh spectral evaluation. The operational matrix $\underline{E}$ and the matrix $\underline{\mathcal{J}}$ should also be multiplied with a factor corresponding to the base width of I_i.

4. Apply BAPILLTIS in each I_i. The identification equation in the ith subinterval will be of the form:

$$\underline{\Phi}_{yxt_i}\, \underline{u}_{abf_i} = \underline{c}_{yi} \,, \qquad (1o.13)$$

where

$$\underline{\Phi}_{yxt_i} = [\underline{\Phi}_{y_i} \,\vdots\, \underline{\Phi}_{xd_i} \,\vdots\, \underline{\mathcal{J}}_i] \,,$$

$$\underline{c}_{y_i} = -[\underline{E}_i^n]^T \underline{Y}_i \,,$$

$$\underline{\Phi}_{y_i} = [\underline{E}_{i\ n-1}^T \underline{Y}_i \,\vdots\, \underline{E}_{i\ n-2}^T \underline{Y}_i \,\vdots \cdots \vdots\, \underline{Y}_i] \,,$$

$$\underline{\Phi}_{xd_i} = -\underline{E}_{i\ n}^T \underline{x}_{d_i}^T \,, \text{ and}$$

$\underline{u}_{abf_i}$ is the vector of unknowns as determined from the data on I_i.

5. Compute the error function

$$H_\tau(\tau) = \sum_{i=1}^{L} \|\underline{u}_{abf_{i+1}} - \underline{u}_{abf_i}\| \,. \qquad (10.14)$$

In a deterministic situation, $H_\tau(\tau)=0$, when τ attains the actual value. The parameters and initial condition terms in $\underline{u}_{abf}$, i.e., $\{a_1, a_2, \ldots, a_n, b, f_1, f_2, \ldots, f_n\}$ should be invariant over all the subintervals. In the presence of noise $H_\tau(\tau)$ attains a minimum at the true value of τ.

6. Minimize $H_\tau(\tau)$ with respect to τ. The parameter vector corresponding to this minimum gives the system parameters.
   To do this, we may use Newton-Raphson method. We choose a small increment $\Delta\tau$ and compute

$$H'_\tau(\tau) = [H_\tau(\tau+\Delta\tau) - H_\tau(\tau)]/\Delta\tau. \tag{10.15}$$

   $H'_\tau(\tau)$ is used to guide the iteration. At each stage the value of τ obtained is used to derive $x_d(t) = x(t-\tau)$. In view of this iterated delay, this algorithm is termed as the iterative shift algorithm (ISA) [W47].

<u>Example 10.5.</u>

Consider a system modelled by

$$\frac{d^2y(t)}{dt^2} + a_1\frac{dy(t)}{dt} + a_o\, y(t) = b\, x(t-\tau) \quad ,$$

where $y(t)$ and $x(t)$ are available in the interval [1.5, 3.0]. We wish to determine a_o, a_1, b, τ and the state $(y, \dot{y})^T$ at the initial instant of reference. The coefficient of $\frac{d^2y}{dt^2}$ is taken as unity without loss of generality. With the first eight Walsh components the loss function $H_\tau(\tau)$ for the system, for $L = 3$, $t_1 = t_o = 1.5$, $t_1 = 2.0$, $t_2 = 2.5$ and $t_3 = 3.0$ sec is shown in Table 10.1. A set of data, from a simulated system with unit ramp input sampled at 16 per sec., is used in the iterative shift algorithm. Table 10.2 shows the pattern of convergence. As and when additional data points are required between the sampled points, linear interpolation is used. A suitable bound may be specified for $H_\tau(\tau)$ and the iteration may be terminated when $H_\tau(\tau)$ has value within this bound.

Table 10.1

Loss function in the Example 10.5

τ	0	1/16	1/8	3/16	1/4	5/16	3/8	7/16	1/2	9/16	5/8	11/16
$H_\tau(\tau)$	0.1772	0.1678	0.1563	0.1419	0.1234	0.0997	0.0692	0.0325	True lag 0.0017	0.0553	0.7151	7.107

Table 10.2.

Pattern of convergence of the iterative shift algorithm (Example 10.5)

Iteration Number	τ_k	$H_\tau(\tau_k)$	$\Delta\tau_k$	$H_\tau(\tau_k+\Delta\tau_k)$	τ_{k+1}	a_1	a_o	b	$y(t_1)$	(t_1)
1	0.15	0.1509	0.3	0.0823	0.5897	4.0415	2.7003	1.3917	0.0838	0.2000
2	0.5897	0.1898	0.1	7.7303	0.5872	4.0021	2.6737	1.3768	0.0838	0.2000
3	0.5872	0.1713	0.05	1.1045	0.5780	3.8654	2.5825	1.3252	0.0838	0.2000
4	0.5780	0.1169	0.025	0.3194	0.5638	3.6727	2.4513	1.2524	0.0838	0.2000
5	0.5638	0.0591	0.0125	0.1073	0.5485	3.4881	2.3265	1.1827	0.0838	0.2000
6	0.5485	0.0245	0.00625	0.0360	0.5352	3.3444	2.2293	1.1284	0.0838	0.2000
7	0.5352	0.0088	0.003125	0.0116	0.5254	3.2459	2.1626	1.0911	0.0838	0.2000
8	0.5254	0.0029	0.0015625	0.0036	0.5186	3.1821	2.1194	1.0670	0.0838	0.1999
9	0.5186	0.0009	0.00078125	0.0011	0.5137	3.1377	2.0893	1.0502	0.0838	0.2000
Actual values used in simulation					0.5	3.0	2.0	1.0	0.0840	0.1997

10.6. Discussion

Walsh spectral treatment has certain inherent filtering features. For instance, if the Walsh spectrum of a signal x(t) is $\{x_i\}$, and if the signal is corrupted with zero mean noise $\zeta_n(t)$, then the Walsh spectrum of $[x(t)+\zeta_n(t)]$ is $\{\hat{x}_i\}$, in which $\hat{x}_i \to x_i$ for all i of interest if the base is sufficiently large. The disparity in the spectral components, $|\hat{x}_i - x_i| \to 0$ in particular, as $i \to 0$. In practical situations where this disparity is not insignificant towards large i, $H_\tau(\tau)$ is likely to have a nonzero minimum at the actual value of lag. In such cases suitbale minimum seeking techniques may be employed. Systems with multiple lags may be handled by searching for the absolute minimum of a suitably defined multivariable error function $H_\tau(\tau_1,\tau_2,\ldots)$. We have deliberately ignored knowledge of y(.) in the initial state and treated the whole set of initial conditions as unknown. This would slightly increase the dimension of the vector $\underline{u}_{abf}$, but prevents any noise in y(t) from directly effecting the matrices $\underline{\Phi}_{yxt}$ and $\underline{c}_y$ which are now wholly in terms of noise resistant Walsh spectral components.

Table 10.3 shows how Walsh spectra are immune to zero mean additive noise. The signal is a unit ramp on the interval [0,1). Over this interval 512 uniformly sampled points of data are considered with different ratios of noise to signal.

Notice that Walsh spectral characterization of the above signal leads to natural filtering of the noise to about 20 % of its original level. The base of Walsh functions may be gradually increased to accomodate additional data as it arrives. This enchances the inherent filtering process further.

Table 10.3. Walsh spectra of a noisy signal, $f(t) = t+\zeta_n(t)$

i	Walsh spectral components, f_i with NSR 0	0.1	0.2
0	0.5	0.49782	0.49691
1	-0.25	-0.25091	-0.25129
2	-0.125	-0.12779	-0.12894
3	0	-0.00338	-0.00478
4	-0.0625	-0.06577	-0.06712
5	0	0.00302	0.00427
6	0	0.00229	0.00323
7	0	-0.00265	-0.00375
8	-0.03125	-0.03284	-0.03350
9	0	-0.03284	-0.00065
10	0	0	0
11	0	0.00612	-0.00866
12	0	0.00809	-0.01144
13	0	0.00704	0.00995
14	0	0.00155	-0.00220
15	0	-0.00119	-0.00168
Noise to signal ratio in Walsh spectral characterization	0	0.01943	0.03796

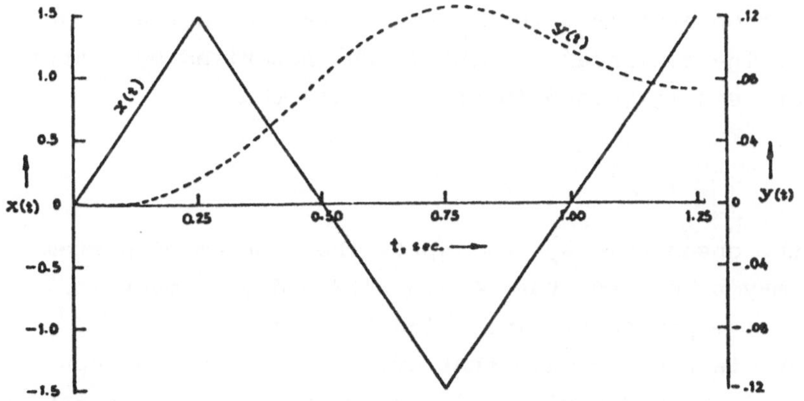

Fig. 10.1. Signals in Example 10.4

10.7. Identification of piecewise linear systems

Nonlinear elements in a process are often modelled by analytic descriptions such as power series, Volterra series, etc. Billings [I3] recently surveyed some important developments in the area of identification of non-linear systems. In spite of their undoubted generality, analytic models have the limitation of being less accurate for systems which are highly nonlinear, such as those exhibiting switching phenomena. For instance, characteristics exhibiting abrupt saturation and dead zone are better modelled by piecewise linear characteristics. Piecewise linear models in such cases have less number of parameters than the corresponding analytic models. Consequently, in algorithms for parameter identification, piecewise linear models are lighter to handle. In view of this situation, we discuss a method of parameter identification in non-linear systems using piecewise linear models. The technique suggested is an iterative level algorithm (ILA). The ILA may be seen to be close in spirit to the iterative shift algorithm (ISA) [W47,I16] for time-lag systems. For the purpose of the present discussion, we consider a class of models of the form shown in Fig. 10.2. The linear element is chosen to be in the general form:

$$y(t) + \sum_{i=1}^{n} a_i \frac{d^i y(t)}{dt^i} = b_o \overset{*}{x}(t) + \sum_{j=1}^{n-1} b_j \frac{d^j \overset{*}{x}(t)}{dt^j}, \qquad (10.16)$$

where $y(t)$ is the observable output and $\overset{*}{x}(t)$ the unobservable input to the linear part. The nonlinear element is characterised by a static, piecewise linear, single valued function $\mathcal{N}$ such that

$$\overset{*}{x} = \mathcal{N}(x),$$

where $x(t)$ is the observable system input. The problem of system identification may be stated thus: Given $x(t)$ and $y(t)$ over an arbitrary but active period of time, find the a's, b's and $\mathcal{N}$. The term 'active' is used to mean that over the record the input is able to excite all the modes of the linear part and scan the essential domain of $\mathcal{N}$. Owing to the arbitrariness of the data record, there are other unknowns:

$$[\alpha_{yi} \triangleq \left.\frac{d^{i-1}y}{dt^{i-1}}\right|_{t=0}, \quad i=1,2,\ldots,n-1] \quad \text{and}$$

$$[\beta_{\overset{*}{x}j} = \left.\frac{d^{j-1}\overset{*}{x}}{dt^{j-1}}\right|_{t=0}, \quad j=2,3,\ldots,n-1] .$$

Although y(0) is actually known, we will ignore this information and treat it as unknown along with the other initial conditions. As usual we arrange the equations in the form (10.2), in which for the present situation $\underline{\phi}_x$ is replaced by $\underline{\phi}_{\overset{*}{x}}$ and:

$$\underline{\Phi}_{\overset{*}{x}} = -[\underline{E}_n^T\overset{*}{\underline{x}} \;|\; \underline{E}_{n-1}^T\overset{*}{\underline{x}} \;|\; \ldots \;|\; \underline{E}^T\overset{*}{\underline{x}}] , \tag{10.17a}$$

$$\overset{*}{x}(t) = \mathcal{N}[x(t)] = \overset{*}{\underline{x}}^T \underline{w}(t) , \tag{10.17b}$$

$$\overset{*}{\underline{x}} = [\overset{*}{x}_0, \overset{*}{x}_1, \ldots, \overset{*}{x}_{m-1}]^T , \tag{10.17c}$$

$$\overset{*}{x}_i = \int_0^1 \mathcal{N}[x(t)]w_i(t)dt . \tag{10.17d}$$

Let $\mathcal{N}$ be piecewise linear with a set of break points $^N P$: $[s_i, m_i, i=1,2,\ldots,N]$. The proposed algorithm consists of the following steps:

Step 1: For a chosen set of break points $^N P$ on $\mathcal{N}$, process x(t) available on $(0,T_0)$ to get $\overset{*}{x}(t)$.

Step 2: Consider subintervals $I_i,(0,T_i)$, $T_i \in T$, $i=1,2,\ldots,L$, and obtain Walsh spectral expansions $\underline{y}$ and $\overset{*}{\underline{x}}$ on each subinterval I_i.

Step 3: Form equation (10.2) and solve for $\underline{u}_{abf}$ the set of unknowns separately on the basis on $\underline{y}$ and $\overset{*}{\underline{x}}$ on each subinterval. Let $\underline{u}_{abf_i}$ denote the parameter vector obtained from data on the i-th subinterval.

Step 4: Compute the error function

$$H_{\mathcal{N}}(^N P) = \sum_{i=1}^{L-1} \|\underline{u}_{abf_{i+1}} - \underline{u}_{abf_i}\| . \tag{10.18}$$

Step 5: Minimise $H_{\mathcal{N}}$ with respect to ${}^{N}P$. Let P^{o} represent the optimum set of break points on $\mathcal{N}$. The parameters $[\underline{a}^{T}\ \underline{b}^{T}]^{T}$ corresponding to P^{o} represent the linear element.

The above algorithm has been applied in two situations discussed in the following:

Example 10.6. System with saturation.

A system with the following parameters and conditions has first been simulated
Linear element with transfer function = $1/(s^{2}+s+1)$

Nonlinear element $\mathcal{N}(x) = k_{1}x = 2x, \quad x \leq 0.25$
$= 0.5$ otherwise.

The process data is shown in Fig. 10.3.
The model chosen is of the form:
Linear element: Equation (10.16) with n=2.
Nonlinear element: Piecewise linear with one break point S on x to be determined.
Number of subintvals L=3. Number of Walsh components, m=16. With the simulated data the error function $H_{\mathcal{N}}(s)$ is shown in Fig. 10.4.
The minimum occurs at S=0.25. The corresponding parameters are:
$H_{\mathcal{N}}^{o} = H_{\mathcal{N}}(S=0.25) = 0.01169$, $a_{2} = 1.0143$, $a_{1} = 1.0354$, $b_{o}k_{1} = 2.09$.

Example 10.7.: System with dead zone
The following system has first been simulated with unit ramp input.
Linear element with transfer function = $b/(as+1)$, b=3, a=2.
Nonlinear element with dead zone D=0.75, and gain beyond dead zone, K=2.0. The input-output data obtained from the above simulation has been considered on three arbritrary subintervals: (0.75, 1.75), (1.25, 2.25), (1.75, 2.75). The error function $H_{\mathcal{N}}(D)$ is shown in Fig. 10.5. It has a minimum at D=0.75. At this point $H_{\mathcal{N}}^{o} = H_{\mathcal{N}}(D=0.75) =$ 0.0002, a = 1.9934, bK = 5.98.

Conclusion
A method of identifying piecewise linear systems via Walsh functions has been outlined. An error function $H_{\mathcal{N}}(.)$ is minimised with respect to the coordinates of a set of break points on $\mathcal{N}(.)$. Characterization of $\mathcal{N}$ in terms of break points often involves less number

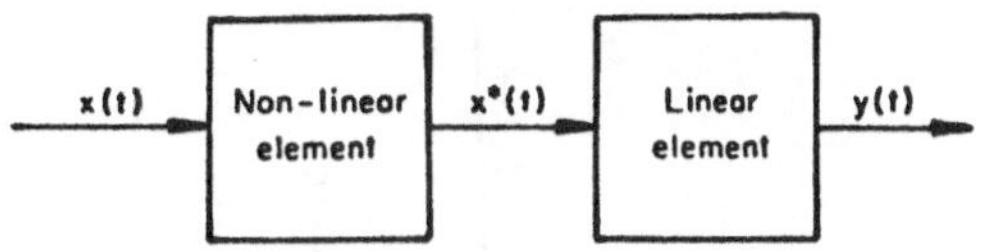

Fig. 10.2. A system with input nonlinearity

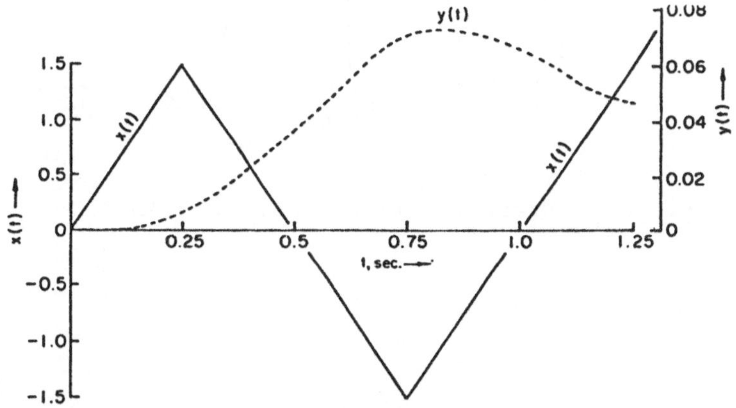

Fig. 10.3. Data for Example 10.6.

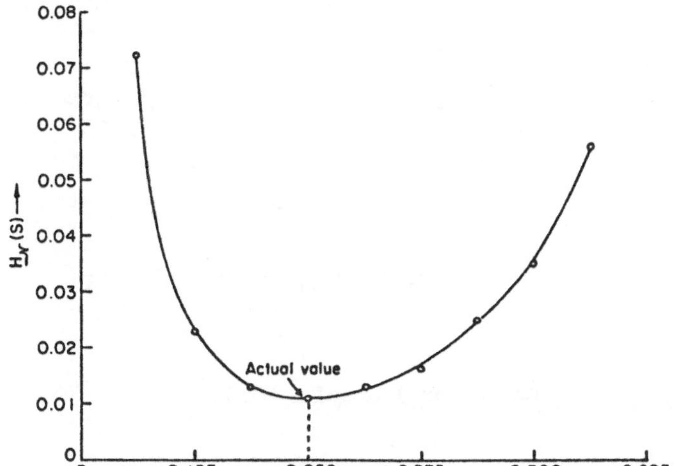

Fig. 10.4. The error function in Example 10.6.

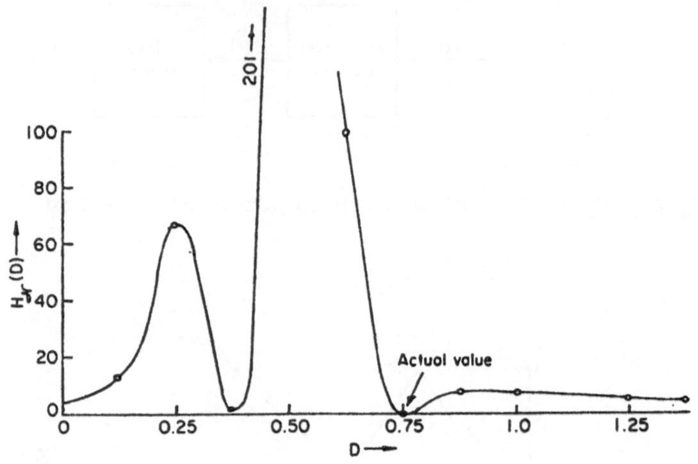

Fig. 10.5. Error function in Example 10.7.

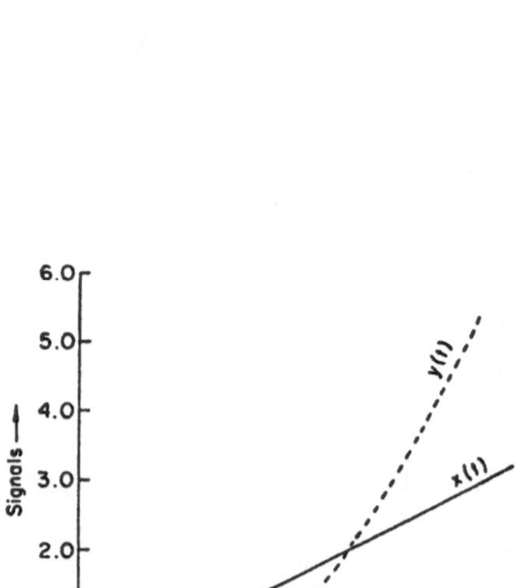

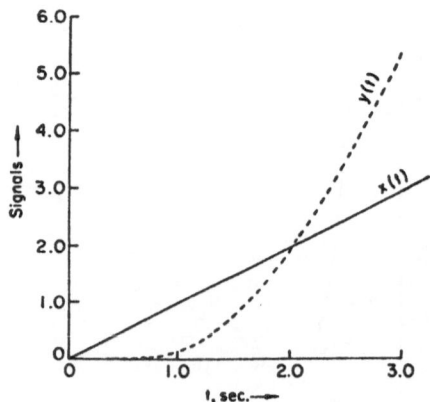

Fig. 10.6. Data for Example 10.7.

of parameters than the conventional power series representation. Consequently, the identification problem is reduced in size leading to certain computational advantages, particularly reduced storage requirements.

Minimization of $H_{\mathcal{N}}$ has to be by an iterative level algorithm (ILA) in which the question of convergence naturally arises. Convergence depends on the nature of $H_{\mathcal{N}}$ which in fact mostly depends on the given data. Therefore, it is not easy to say much about the convergence of the ILA in general. It would be better if one determines the neighbourhood in which $H_{\mathcal{N}}^{o}$ occurs probably by direct search prior to applying any scheme of iteration.

In conclusion, the ILA is recommended in situations where one can afford more time than storage for computation.

10.8. Improved BAPILLTIS using OSOMRI

With the following changes in the existing BAPILLTIS, the accuracy of the identification algorithm is improved in the case of higher order models. Equation (10.3) will be ignored now and the OSOMRI from equation (2.15) will be employed. Correspondingly equation (10.3) will be revised as

$$\underline{\mathcal{J}}_{(n\times m)} = \left[\begin{array}{c} \text{1st row of } \underline{E}^{o}=\underline{I} \\ \hline \text{1st row of } \underline{E} \\ \hline \\ \text{1st row of } \underline{E}_{n-1} \end{array}\right]. \qquad (10.3)$$

The following examples show the superiority of the improved algorithm in higher order models.

Example 10.8.: Consider the problem of identification in a second order lag-free model given by

$$\ddot{y}(t) = a_1 \dot{y}(t) + a_2 y(t) = x(t),$$

with input-output data on [0,1) simulated in two cases with unit step input and zero initial conditions. Case 1: $a_1 = 0.3$, $a_2 = 0.02$

and case 2: $a_1 = 0.12$, $a_2 = 0.002$.
The improved identification algorithm is applied by letting n=2 with a set of 8 Walsh functions (m=8). Results due to $\underline{E}_j$ and $\underline{E}^j$ are shown together against the actual results in Tables 10.4 (a) and 10.4 (b).

Table 10.4. (a)
Identification in Example 10.8. Case 1.

Parameters	a_1	a_2	$y(0)$	$\dot{y}(0)$	Mean square error (MSE)
True values	0.3	0.02	0.0	0.0	-
Result due to $\underline{E}^2$	0.30025	0.019442	0.000774	-0.001302	2.69266×10^{-6}
Result due to $\underline{E}_2$	0.300405	0.019227	0.000380	0.000001	9.05955×10^{-7}

Table 10.4. (b)
Identification in Example 10.8. Case 2.

Parameters	a_1	a_2	$y(0)$	$\dot{y}(0)$	Mean square error (MSE)
True values	0.12	0.002	0.0	0.0	-
Result due to $\underline{E}^2$	0.120061	0.001881	0.000311	-0.001302	1.8098×10^{-6}
Result due to $\underline{E}_2$	0.120072	0.001860	0.000154	0.000000	4.85×10^{-8}

The Mean Square Error (MSE) is $\sum_i (u_i-\bar{u}_i)^2$, where u_i und $\bar{u}_i$ denote respectively, the true and the calculated values of system parameters. This is a reasonable basis for comparison of the results due to $\underline{E}_j$ and $\underline{E}^j$ with the actual ones. In the present case $\underline{E}_2$ clearly gives better resuts than $\underline{E}^2$.

10.9. Identification in the presence of small unknown time-delays

When the model happens to contain delay terms, the iterative shift algorithm (ISA) seems to be the only way to handle the situation. In this section we will see how the Walsh Delay Matrices (WDM) render the algorithm direct for small delays. The ISA which is inevitable for large delays may be generally improved by using WDM. In order to understand the non-iterative method for small delays, let us consider the model

$$y(t) + \sum_{i=1}^{n} a_i \frac{d^i y(t)}{dt^i} = b_o\, x(t-\tau_x)\,, \tag{10.19}$$

with an unknown small delay τ_x in the input process. We now integrate (10.19) n times and insert the operational matrices $\underline{E}_j$. The delay term

$$x(t-\tau_x) = \underline{x}^T \underline{D}_{(m)}^{\varepsilon_x} \underline{w}(t) = [\underline{x}^T + \varepsilon_x \underline{x}^T \underline{\Phi}_{d_{(m)}}]\underline{w}(t)\,, \tag{10.20}$$

where

$$\tau_x = \frac{\varepsilon_x}{m} \quad \text{and } \underline{\Phi}_{d_{(m)}} \quad \text{is as in (2.29).}$$

This leads to

$$[\underline{\Phi}_y \;\vdots\; -\underline{E}_n^T \underline{x} \;\vdots\; -(\underline{\Phi}_{d_{(m)}} \underline{E}_n)^T \underline{x} \;\vdots\; \underline{\mathcal{J}}^T]\underline{u}_{ab\varepsilon f} = \underline{c}_y, \tag{10.21}$$

where

$$\underline{u}_{ab\varepsilon f} = [a_1,\, a_2, \ldots, a_n,\, b_o,\, \varepsilon_x b_o,\, f_1,\, f_2, \ldots, f_n]^T.$$

Equation (10.21) gives the parameters and the delay in terms of ε_x.

Example 10.9.: Consider a process described by

$$\dot{y}(t) + a_1 y(t) = x(t-\tau_x), \tag{10.22}$$

with a small unknown delay τ_x in the input process. Data generated with zero initial conditions and unit step input is employed in the algorithm.

Table 10.5 summarizes the results obtained in different cases of numerical experiment. Notice how the unknown inital condition also is obtained along with the other parameters.

Table 10.5.

Identification of parameters in a first order model (Example 10.9)

No.		Parameters			
		a	y(0)	Delay	ε_x
1.	True values	0.5	0.0	0.0292968	0.9375
	Result	-0.510261	-0.000388	0.0298809	0.956189
2.	True values	-1.0	0.0	0.0292968	0.9375
	Result	-1.006355	-0.001077	0.0287801	0.920966
3.	True values	-2.0	1.0	0.0273437	0.875
	Result	-1.99977	0.99735	0.0248665	0.795731
4.	True values	-2.0	0.0	0.0292968	0.9375
	Result	-2.000475	-0.001772	0.027408	0.877059

Example 10.10.

Consider a second order model of the form

$$\ddot{y}(t) + a_1\dot{y}(t) + a_2y(t) = x(t-\tau_x), \tag{10.23}$$

with $\tau_x = -\frac{\varepsilon_x}{m}$, $0 \le \varepsilon_x \le 1$. Data generated with zero initial conditions and unit step input is employed in the algorithm and the results are summarized in Table 10.6. Notice the automatically evaluated initial conditions and the superiotity of E_2.

10.10. Identification of systems with larger time-delays

When τ_x is not small, expressing it as fraction of $\frac{1}{m}$ will require small m. This will influence the numerical accuracy of representation and leads to large errors in the results. We should therefore let m remain as large required and correspondingly modify our approach.

Table 10.6.

Identification of parameters in a second order model (Example 10.10).

No.	Parameters	a_1	a_2	ε_x	$\dot{x}(0)$	$x(0)$	MSE $\sum_i (u_i - \bar{u}_i)^2$
1	True values	0.3	0.02	0.875	0.0	0.0	-
	Result using $\underline{E}_2$	0.299940	0.02139	0.675562	-0.006217	0.000045	0.0398159
	Result using $\underline{E}^2$	0.299924	0.021395	0.450359	-0.013258	0.000073	0.1804977
2	True values	0.25	0.01	0.875	0.0	0.0	-
	Result using $\underline{E}_2$	0.249953	0.01131	0.675716	-0.006214	0.000045	0.0397544
	Result using $\underline{E}^2$	0.249944	0.011346	0.450469	-0.013257	0.000073	0.1804041

In what follows, we present such an algorithm. The delayed term is written as

$$x(t-\tau_x) = \underline{x}^T \underline{D}_{(m)}^{N+\varepsilon_x} \underline{w}(t) = [\underline{x}^T\underline{D}_{(m)}^{N} + \varepsilon_x\underline{x}^T \underline{\Phi}_{d(m)}\underline{D}_{(m)}^{N}]\underline{w}(t),$$

where

$$\tau_x = \left(\frac{N+\varepsilon_x}{m}\right) , \; 0\leq\varepsilon_x\leq 1.$$

We will at first, try to determine N and then apply the direct algorithm to find ε_x. The form of the identification equation for (10.19) is the same as (10.21) with a difference. That is, the term $-(\underline{E}_n \underline{D}_{(m)}^N)\underline{x}$ will appear in place of $-(\underline{\Phi}_{d(m)}\underline{E}_n)^T\underline{x}$. Given the input-output record over $[t_o,t_f)$, the algorithm may be applied in the following steps.

i) Choose WF defined on the base $[t_1, t_{d1J}, t_1, t_d \in [t_o,t_f)$

ii) Form equation (10.2) with $\underline{D}_{(m)}^N$ in place of $\underline{D}_{(m)}^{\varepsilon_x}$

iii) If m>2n+1, the number of parameters, apply the principle of least squares and normalize (10.2).

iv) Calculate the parameter vector $\underline{u}_1$

v) Repeat the steps (i) to (v) choosing a base (t_1,t_{d2}) for WF where $t_1, t_{d2} \in [t_o,t_f)$ and obtain the corresponding parameter vector $\underline{u}_2$.

vii) Compute the disparity between $\underline{u}_1$ and $\underline{u}_2$ as

$$H_d(N) = ||\underline{u}_1-\underline{u}_2||$$

For a value of N corresponding to the neighbourhood of τ_x, $H_d(N)$ becomes minimum.

viii) Obtain N by direct search

ix) Employing the derived value of N compute ε_x following the direct algorithm on the lines of Section 10.9.

Example 10.11. Consider a delay process

$$\dot{y}(t) = a\, y(t) + x(t-\tau_x), \tag{10.24}$$

with a unit step input. The process is simulated with a=-0.5, y(0) = 1, and τ_x = 0.0527343. For the chosen value of 1/m i.e. 1/16, ε_x = 0.84375. The function $H_d(N)$ is shown in Fig. 10.7. It may be seen that N=0 has the least $H_d(N)$. This means that the actual value of lag lies between N=0 and 1. Now we apply the direct algorithm and get the parameters including ε_x. The final results are shown on Table 10.7. The second set of the results in Table 10.7 belongs to another situation in which the actual delay is fairly large such that for N=4, $H_d(N)$ is minimum. The function $H_d(N)$ is shown in Fig. 10.8.

10.11. Identification in the presence of several small delays

Consider a system with two delays for an illustrative development

$$y(t-\tau_y) + \sum_{i=1}^{n} a_i \frac{d^i y(t)}{dt^i} = b_o x(t-\tau_x), \tag{10.25}$$

where τ_x and τ_y are small. We will write

$$y(t-\tau_y) = \underline{y}^T \underline{D}_{(m)}^{\varepsilon_y} \underline{w}(t) = [\underline{y}^T + \varepsilon_y \underline{y}^T \underline{\Phi}_{d\,(m)}]\, \underline{w}(t),$$

and

$$x(t-\tau_x) = \underline{x}^T \underline{D}_{(m)}^{\varepsilon_x} \underline{w}(t) = [\underline{x}^T + \varepsilon_x \underline{x}^T \underline{\Phi}_{d\,(m)}]\underline{w}(t) .$$

The identification equation then becomes

$$[\underline{\Phi}_y \mid (\underline{\Phi}_{d(m)}\, \underline{E}_n)^T \underline{y} \mid -\underline{E}_n^T\, \underline{x} \mid -(\Phi_{d(m)} \underline{E}_n)^T \underline{x} \mid \mathcal{J}^T]\; \underline{u}_{abf\varepsilon_x\varepsilon_y} \tag{10.26}$$

$$= \underline{c}_y ,$$

where $\underline{u}_{ab\varepsilon_x\varepsilon_y} = [a_1, a_2, \ldots, a_n, \varepsilon_y, b_o, \varepsilon_x, b_o]^T$.

Table 10.7.

Results of multistep algorithm for system suspected to have large delays

No	Parameters	a	y(O)	ε_x	N	τ_x
1	True values	-0.5	1.0	0.84375	0	0.0527343
	Result	-0.500082	0.991784	0.710999	0	0.044434
2	True values	-0.5	1.0	0.84375	4	0.5527343
	Result	-0.500264	0.000750	0.839935	4	0.5524959

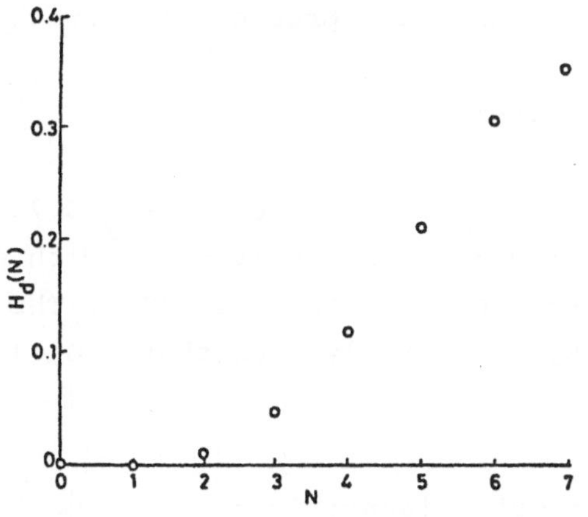

Fig. 10.7. Error function in Example 10.11. (m=16)

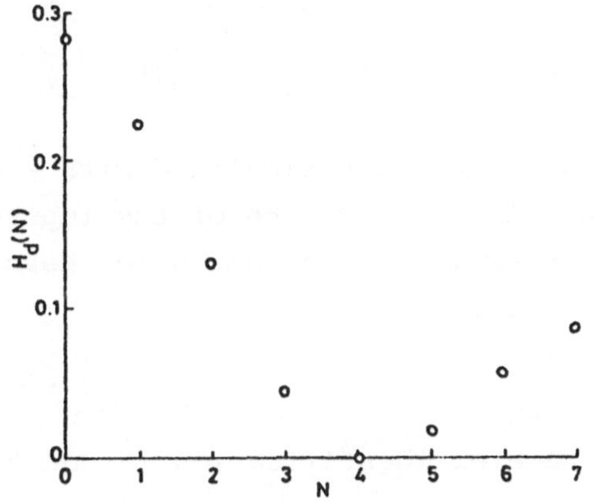

Fig. 10.8. Error function in Example 10.11. (m=16)

Example 10.12. Consider a process with two delay terms:

$$\dot{y}(t) = a\, y(t-\tau_y) + x(t-\tau_x) . \qquad (10.27)$$

With a=-0.5, $\tau_x=\tau_y=0.046875$ i.e., $\varepsilon_x=\varepsilon_y=0.75$. The input-output data is generated with x(t) as a unit step. With m=16 the results of computation are presented in Table 10.8. The data is used from the record begining arbitrarily such that y(0)=1.0.

Table 10.8.

Results of parameter identification in equation (10.27)

Para-meters	a	y(0)	ε_x	ε_y	τ_x	τ_y
True values	-0.5	1.0	0.75	0.75	0.046875	0.046875
Result	-0.500283	0.994009	0.663679	0.773617	0.414799	0.048351

Example 10.13. Consider a process with two delayed inputs each with a small delay

$$\dot{y}(t) = a\, y(t) + x_1(t-\tau_{x1}) + x_2(t-\tau_{x2}) . \qquad (10.28)$$

Input-output data has been simulated with a=0.5, $\tau_{x1}=0.0234375$, $\tau_{x_2}=0.0292968$ and y(0)=0.0. One of the inputs is a delayed unit ramp and the other a delayed unit step. Results of computation are shown in Table 10.9.

Table 10.9.

Results of parameter identification in equation (10.28)

Para-meters	a	y(0)	ε_{x1}	ε_{x2}	τ_{x1}	τ_{x2}
True values	-0.5	0.0	0.75	0.9375	0.0234375	0.0292968
Result	-0.501620	-0.001635	0.884831	0.944916	0.0276509	0.0296536

The error in Walsh series approximation of repeated integrals is limited to only one stage as a result of the newly proposed one shot operational matrices for repeated integration. The algebraically simple idea of using higher powers of $\underline{E}$ results in an accumulation of errors at each stage of integration. Moreover generation of $\underline{E}^j$ from the basic $\underline{E}$ matrix by successive matrix multiplications is computationally less economical than the generation of $\underline{E}_j$ in which $\underline{\Delta}^r$, the power of the shift matrix can be effected by simply shifting the diagonal of unity elements appropriately. The superiority of the operational matrix $\underline{E}_j$ is clearly demonstrated in algorithms of parameter identification. A direct non-iterative algorithm is suggested for systems containing small time delays. A multistep algorithm is proposed for problems involving large delays.

10.12. Parameter identification in bilinear, time-varying, and nonlinear systems

When parameter identification in bilinear, time-varying and non-linear models is carried out we come across certain terms whose PCBF approximations should first be established. The general forms of such terms and their BPF expansions, which are the simplest, are discussed below. The subscript β is dropped for convenience.

i) If $f_i(t)$, $i=0,1,\ldots,n$, $t\in[0,1)$ are square integrable functions and if the BPF expansions are

$$f_i(t) \simeq \underline{f}_i^T \ \underline{\beta}(t), \ i=0,1,2,\ldots,n \ , \tag{10.29a}$$

where

$$\underline{f}_i = (f_{i1} \ \ f_{i2} \ \ \ldots \ f_{im})^T, \tag{10.29b}$$

then, the BPF expansion of the product of the functions is given by

$$\prod_{i=1}^{n} f_i(t) \approx \underline{\underline{f}}^T \ \underline{\beta}(t) \ , \tag{10.29c}$$

where

$$\underline{\underline{f}} = (\prod_{i=1}^{n} f_{i1} \ \ \prod_{i=1}^{n} f_{i2} \ \ldots \ \prod_{i=1}^{n} f_{im})^T . \tag{10.29c}$$

ii) If

$$f(t) = (f_1 \ \ f_2 \ \ldots \ f_m)^T \ , \tag{10.30a}$$

and if

$$g(t) = \mathcal{N}(f(t)) \ , \tag{10.30b}$$

where $\mathcal{N}(.)$ is a function, which may be nonlinear, then

$$g(t) \simeq (\mathcal{N}(f_1) \ \ \mathcal{N}(f_2) \ \ldots \mathcal{N}(f_m))^T \ \underline{\beta}(t) \ . \tag{10.30c}$$

iii) The integral

$$\int_o^t \prod_{i=1}^n f_i(t)dt \tag{10.31a}$$

may be expressed as

$$\begin{bmatrix} \frac{1}{2}\prod_{i=1}^n f_{i1} & \prod_{i=1}^n f_{i1} & \prod_{i=1}^n f_{i1} & \cdots & \prod_{i=1}^n f_{i1} \\ & \frac{1}{2}\prod_{i=1}^n f_{i2} & \prod_{i=1}^n f_{i2} & \cdots & \prod_{i=1}^n f_{i2} \\ & & \cdots\cdots\cdots & & \\ o & & & & \frac{1}{2}\prod_{i=1}^n f_{im} \end{bmatrix} . \tag{10.31b}$$

The above relations will be instrumental in decomposing the calculus of continuous time-varying, nonlinear and bilinear models into an algebra approximate in the sense of least squares. Karanam et al. [W30] and Jan and Wang [W28] gave identification algorithms for bilinear models via WF and BPF respectively. The relations may be derived with any PCBF as the basis, but, it is through BPF they appear in the simplest form as given above. We will therefore discuss the problems in the sequel in the BPF domain only.

<u>Bilinear models:</u> Usually a bilinear model has the form:

$$\dot{\underline{x}}(t) = \underline{A}(t)\underline{x}(t) + \sum_{i=1}^k \underline{C}_i(t)\underline{x}(t)\underline{u}_i(t) + \underline{B}(t)\underline{u}(t), t \geq o . \tag{10.32a}$$

The output is given by

$$\underline{y}(t) = d(t)\underline{x}(t) , t \geq o . \tag{10.32b}$$

$\underline{u}(t)$ is a $\underline{k}$-dimensional real vector, $\underline{x}$ and $\underline{y}$ are n dimensional and $\underline{A}$, $\underline{B}$ and $\underline{C}$ are of appropriate dimensions. y_i, u_i are the i-th component of $\underline{y}$ and $\underline{u}$ respectively.

The input-output relationship may be expressed as

$$\underline{y}(t) = \underline{y}(0) + \int_o^t \sum_{j=1}^n \underline{a}_j(\tau)y_i(\tau)d\tau + \int_o^t \sum_{i=1}^k \sum_{j=1}^n \underline{c}_{ij}(\tau)y_i(\tau)d\tau$$
$$+ \int_o^t \sum_{i=1}^k \underline{b}_i(\tau)d\tau\underline{u}_i(\tau)d\tau + \int_o^t \frac{\dot{d}}{d}(\tau)\underline{y}(\tau)d\tau , \tag{10.33}$$

where $\underline{a}_j(t)$, $\underline{c}_{ij}(t)$ and $\underline{b}_j(t)$ are the j-th columns of $\underline{A}(t)$, $\underline{C}_i(t)$ and $\underline{B}(t)$ respectively. The identification problem is now to determine $\underline{a}_j(t)$, $\underline{c}_{ij}(t)$ and $\underline{b}_i(t)$ from a knowledge of $\underline{u}$ and $\underline{y}$.
We now expand the various terms of equation (10.33) in BPF series each of m terms:

$$\underline{a}_j(t) = \underline{A}_j\ \underline{\beta}(t),$$
$$\underline{b}_i(t) = \underline{B}_i\ \underline{\beta}(t),$$
$$\underline{c}_{ij}(t) = \underline{\Gamma}_{ij}\underline{\beta}(t).$$

In these $\underline{A}_j$, $\underline{B}_i$ and $\underline{\Gamma}_{ij}$ are BPF spectral matrices each of m columns.

$$u_i(t) = \underline{u}_i^T\underline{\beta}(t), \tag{10.34}$$

$$y_j(t) = \underline{y}_i^T\underline{\beta}(t),$$

$$d(t) = \underline{d}^T\underline{\beta}(t),$$

and

$$\frac{d}{d}(t) = \underline{d}_d^T\underline{\beta}(t),$$

where $\underline{u}_i$ $\underline{y}_i$ $\underline{d}$ and $\underline{d}_d$ are m-vectors.

We insert the above in (10.33). The unknowns such as $\underline{A}_j$, $\underline{B}_j$ and $\underline{\Gamma}_{ij}$ are isolated from the part of each term in the RHS of (10.33), known from data. The known parts of the terms involving products and their integrals are evaluated using equations (10.29-31). The result is a set of equations linear in the unknowns which can be solved easily. The relations (10.29-31) and their effects in (10.33) become the simplest in the case of BPF. Equation (10.33) with the BPF expansions becomes

$$\begin{aligned}
\underline{Y}\ \underline{\beta}(t) = \underline{y}(0) &+ \int_0^t \sum_{i=1}^{n} \underline{A}_j\ \underline{\beta}(\tau)\underline{\beta}^T(\tau)d\tau \\
&+ \int_0^t \sum_{i=1}^{k} \sum_{j=1}^{n} \underline{\Gamma}_{ij}\underline{\beta}(\tau)\underline{\beta}^T(\tau)\underline{y}_j\underline{u}_i^T\underline{\beta}(\tau)d\tau \\
&+ \int_0^t \sum_{i=1}^{k} \underline{B}_i\underline{\beta}(\tau)\underline{\beta}^T(\tau)\underline{d}\ \underline{u}_i^T\underline{\beta}(\tau)d\tau \\
&+ \int_0^t \underline{Y}\ \underline{\beta}(\tau)\underline{\beta}^T(\tau)\underline{d}_d d\tau\ .
\end{aligned} \tag{10.35}$$

Example 10.14.

Consider the scalar time-invariant system

$$\dot{y}(t) = a\ y(t) + b\ y^2(t)u(t) + c\ y(t)u^2(t) + d\ u(t), \qquad (10.36)$$

$$0 \le t \le 1 .$$

With a=1, b=0.5, c=0.3, d=1.5 and u=exp(-0.5t) and zero initial conditions, the process simulated. The solution y(t) has been obtained at t=eT; e=0,1,...,8; T=0.125.

With BPF expansions we get an equation of the type of (10.35) as follows. The result is a relation given by

$$\underline{y}^T = a\ \underline{y}^T \underline{E}_\beta + \frac{b}{m}\ \underline{y}^T \begin{bmatrix} \frac{y_1u_1}{2} & y_1u_1 & \cdots & y_1u_1 \\ & \frac{y_1u_2}{2} & \cdots & y_2u_2 \\ & & \ddots & \frac{y_mu_m}{2} \end{bmatrix}$$

$$+ \frac{c}{m}\ \underline{y}^T \begin{bmatrix} \frac{u_1^2}{2} & u_1^2 & \cdots & u_1^2 \\ & \frac{u_2^2}{2} & \cdots & u_2^2 \\ & & & \frac{u_m^2}{2} \end{bmatrix} + d\ \underline{u}^T \underline{E}_\beta . \qquad (10.37)$$

Table 10.10 shows the results of computation in two cases.

Table 10.10. Parameter identification in (10.36)

Parameter	True value	Results with BPF m=4	m=8
a	-1.0	-0.9984	-0.9995
b	0.5	0.4964	0.4989
c	0.3	0.2986	0.2986
d	1.5	1.4952	1.4988

Example 10.15.

Consider the scalar time varying bilinear system

$$\dot{y}(t) = a(t)y(t) + c(t)y(t)u(t) + b(t)u(t), \qquad (10.38)$$

With

$$a(t) = -e^{-0.5t},$$

$$b(t) = 1.5\, e^{-0.3t}, \text{ and}$$

$$c(t) = 2t,$$

the process is simulated with four different inputs to generate the required number of equations. With an m-set expansion these 4m equations can be generated to solve for 3m unknowns. The different inputs are as follows:

$$e^{-2t}(0.2), \quad 2t(0.1), \quad te^{-t}(4.1) \text{ and } \frac{1}{t+0.5}\ (2.6).$$

The bracketed values correspond to the initial conditions. Fig. 10.9 shows the results of identification.

Example 10.16.

Consider the second order time-invariant bilinear system modelled by

$$\dot{\underline{y}}(t) = \underline{A}\,\underline{y}(t) + \sum_{i=1}^{2} \underline{C}_i \underline{y}(t) \underline{u}_i(t) + \underline{B}\,u(t). \tag{10.39}$$

With a fourth order Runge-Kutta-method the system is simulated under the following conditions:

$$A = \begin{bmatrix} 0 & 1 \\ -0.5 & -1 \end{bmatrix}, \quad B = \begin{bmatrix} 2 & 1 \\ 1 & 1 \end{bmatrix},$$

$$\underline{C}_1 = \begin{bmatrix} 0 & 1 \\ 0.5 & 0.2 \end{bmatrix}, \quad \underline{C}_2 = \begin{bmatrix} 1 & 0 \\ 0.2 & 0.3 \end{bmatrix},$$

and $\underline{u}(1) = [u_1(t),\ u_2(t)]^T$,

where $u_1(t) = e^{-0.5t}$, and
$u_2(t) = e^{-t}$.

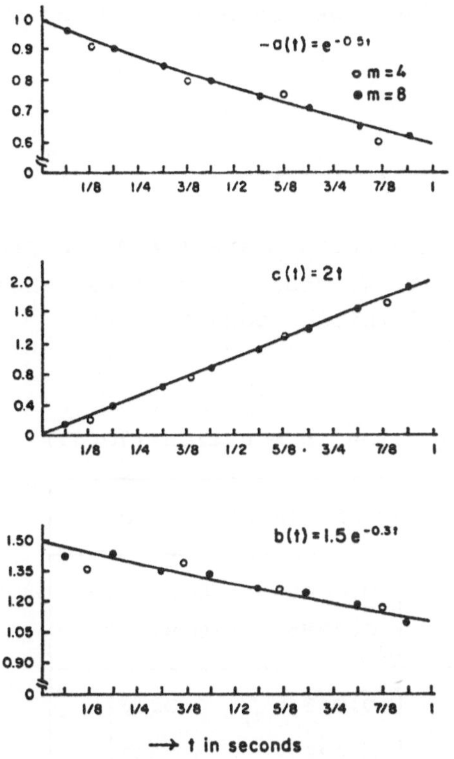

Fig. 10.9. Identification of time-varying parameters in equation (10.38)

The data is obtained from the above at t=l(0.125); l=0,1,...,8. Equation (10.35) in this situation attains the following form:

$$\underline{y}_i^T\underline{\beta}(t) = \underline{y}_i(0) + \sum_{j=1}^{2} a_{ij}\underline{y}_j^T\,\underline{E}_\beta\underline{\beta}(t)$$

$$+ \sum_{j=1}^{2} \underline{c}_{ij}^{(1)}\underline{y}_j^T\,\Big\{\int_0^t \begin{bmatrix} \beta_1(t) & & & \\ & \beta_2(t) & & \\ & & \ddots & \\ & & & \beta_m(t) \end{bmatrix} dt\Big\}u_1$$

$$+ \sum_{j=1}^{2} \underline{c}_{ij}^{(2)}\;\underline{y}_j^T\Big\{\int_0^t \begin{bmatrix} \beta_1(t) & & & \\ & \beta_2(t) & & \\ & & \ddots & \\ & & & \beta_m(t) \end{bmatrix} dt\Big\}u_2$$

$$+ \sum_{j=1}^{2} b_{ij}\underline{u}_j^T\,\underline{E}_\beta\underline{\beta}(t) \qquad i=1,2\ .$$

There are sixteen unknowns. Choosing m=8 and two sets of initial conditions

$$\underline{y}(0) = [0 \quad 0]^T, \quad \text{and } \underline{y} = [0.2 \quad 0.3]^T,$$

sixteen independent equations are formed to yield the results as shown in Table 10.11. We could have chosen alternatively m=16 with only one set of initial conditions.

Table 10.11.
Parameter identification in equation (10.39)

True values	By BPF (m=8) method
$\underline{A} = \begin{bmatrix} 0 & 1 \\ -0.5 & -1 \end{bmatrix}$	$\begin{bmatrix} -0.0046 & 0.9837 \\ -0.4985 & -0.9956 \end{bmatrix}$
$\underline{C}_1 = \begin{bmatrix} 0 & 1 \\ 0.5 & 0.2 \end{bmatrix}$	$\begin{bmatrix} 0.0134 & 1.0246 \\ 0.4967 & 0.1953 \end{bmatrix}$
$\underline{C}_2 = \begin{bmatrix} 1 & 0 \\ 0.2 & 0.3 \end{bmatrix}$	$\begin{bmatrix} 0.9929 & -0.0150 \\ 0.1998 & 0.3014 \end{bmatrix}$
$\underline{B} = \begin{bmatrix} 2 & 1 \\ 1 & 1 \end{bmatrix}$	$\begin{bmatrix} 2.0230 & 0.9680 \\ 0.9925 & 1.0030 \end{bmatrix}$

XI

PARAMETER IDENTIFICATION IN DISTRIBUTED SYSTEMS

Given the output $u(x,t)$ and the distributed input $f(x,t)$ in a two dimensional distributed parameter system (DPS) modelled by a partial differential equation (PDE) which is linear in its constant parameters, over a region $x \in [0,x_o]$, $t \in [0,t_o]$, find the parameters of the system. This is the problem of interest to control engineers. It is the inverse of the problem discussed in chapter IX. We will consider parameter identification in systems modelled by the three partial differential equations treated in chapter IX. The respective identification algorithms may be derived straightaway. Our present treatment will be with two dimensional BPF as the basis without loss of generality with the same nomenclature as in chapter IX.

Example 11.1:

Consider a DPS modelled by the PDE (9.2)

$$\alpha \frac{\partial u(x,t)}{\partial x} + \frac{\partial u(x,t)}{\partial t} = 0.$$

Let $u(x,t)$ be given in a region $x \in [0,1]$, $t \in [0,1]$. By suitable change of scales the region may be normalized if necessary. Our problem is to find α.

Following the definitions and notation in (9.1), let us recall (9.7) from which it is evident that

$$\alpha(\underline{\underline{u}}^T - \underline{\underline{f}}^T)\ \underline{\underline{E}}_t = (\underline{\underline{g}}^T - \underline{\underline{u}}^T)\ \underline{\underline{E}}_x\ ,$$

giving

$$\alpha = (\underline{\underline{g}}^T - \underline{\underline{u}}^T)\ \underline{\underline{E}}_x\ [(\underline{\underline{u}}^T - \underline{\underline{f}}^T)\ \underline{\underline{E}}_t]^{-1}. \qquad (11.1)$$

From the given data u(x,t) the two-dimensional BPF spectrum $\underline{\underline{u}}$, and the spectra of boundary functions $\underline{\underline{f}}$ and $\underline{\underline{g}}$ can be computed using (1.28). The operational matrices $\underline{\underline{E}}_x$ and $\underline{\underline{E}}_t$ are given by

$$\underline{\underline{E}}_x = \underline{\underline{E}}_{\beta\,(m_x \times m_x)} \otimes \underline{\underline{I}}_{(m_t \times m_t)}, \text{ and}$$

$$\underline{\underline{E}}_t = \underline{\underline{I}}_{(m_x \times m_x)} \otimes \underline{\underline{E}}_{\beta\,(m_t \times m_t)}\ ,$$

where m_x and m_t are segments on the unit interval in x and t respectively. Inserting these in (11.1) we get α^2.

<u>Example 11.2:</u>

Consider a DPS modelled by the diffusion equation as in (9.19)

$$\frac{\partial u(x,t)}{\partial t} = \alpha^2\ \frac{\partial^2 u(x,t)}{\partial x^2}\ .$$

Given u(x,t) in $x \in [0,1]$, $t \in [0,1]$, find α^2. Now recall (9.22) which implies that

$$(\underline{\underline{u}}^T - \underline{\underline{g}}^T)\ \underline{\underline{E}}_x^2 = \alpha^2(\underline{\underline{u}}^T - \underline{\underline{f}}^T - \underline{\underline{h}}^T\ E_x)\ \underline{\underline{E}}_t\ ,$$

giving

$$\alpha^2 = (\underline{\underline{u}}^T - \underline{\underline{g}}^T)\ \underline{\underline{E}}_x^2\ [(\underline{\underline{u}}^T - \underline{\underline{f}}^T - \underline{\underline{h}}^T\ \underline{\underline{E}}_x)\ \underline{\underline{E}}_t]^{-1}. \qquad (11.2)$$

In (11.2) $\underline{\underline{u}}$, $\underline{\underline{f}}$ and $\underline{\underline{g}}$ may be directly computed from (1.28). The vector of BPF spectral components of the boundary derivative term $\underline{\underline{h}}$ is however not readily available. This has to be estimated from $\underline{\underline{u}}$ with the help of the partial derivative operator. Such operators as $\underline{\underline{E}}_x^{-1}$ and $\underline{\underline{E}}_t^{-1}$ are evident from our previous discussions. With all the quantities on the RHS of (11.2) so employed, we get α^2.

Example 11.3:

Finally consider a DPS modelled by the wave equation as in (9.31)

$$\frac{\partial^2 u(x,t)}{\partial t^2} = \alpha^2\ \frac{\partial^2 u(u,t)}{\partial x^2}.$$

Given u(x,t) in the unit region the value of α^2 should be estimated. The vectors $\underline{\underline{h}}$ and $\underline{\underline{q}}$ defined in (9.1) may be estimated using the partial derivative operators such as $\underline{\underline{E}}_x^{-1}$ and $\underline{\underline{E}}_t^{-1}$. Recalling (9.33b) we get

$$\alpha^2 = (\underline{\underline{u}}^T - \underline{\underline{g}}^T - \underline{\underline{q}}^T\ \underline{\underline{E}}_t)\ \underline{\underline{E}}_x^2\ [(\underline{\underline{u}}^T - \underline{\underline{f}}^T - \underline{\underline{h}}^T\ \underline{\underline{E}}_x)\ \underline{\underline{E}}_t^2]^{-1}. \qquad (11.3)$$

Use of other PCBF will be on similar lines. The results will be the same. Very little work has so far been reported on this problem. Tzafestas [W70] and Sinha et al. [W64] are the only instances of work in this area. This is perhaps due to the inherent instability of solutions discovered by Prasada Rao and Srinivasan [W55] in problems higher in order than one.

REFERENCES

Optimal Control of time delay systems

C1 J.K. Aggarwal: Computation of optimal control for time delay systems. IEEE Trans. Automatic Contr., Vol 15, pp. 683-685, 1972

C2 J.K. Aggarwal: Feedback control of linear systems with distributed delay. Automatica, Vol. 9, pp. 367-379, 1973.

C3 Y. Alekal, P. Brunovsky, D.H. Chung and E.B. Lee: The quadratic problem for system with time delays. IEEE Trans. on Automatic Control, Vol. AC-16, No. 6, pp. 673-697, 1971.

C4 H.T. Banks: Approximation of Non-linear Functional differential equation control systems. J. Optimization Theory and Applic., Vol. 29, No. 3, pp. 383-408, 1979.

C5 H.T. Banks and J.A. Burns: An abstract framework for approximate solutions to optimal control problems governed by hereditary systems. Proc. of International Conf. on Differential Equations, H.A. Antosiewicz, ed. Academic Press, New York, pp. 10-25, 1975.

C6 H.T. Banks, J.A. Burns, E.M. Clitt and P.R. Thrift: Numerical solutions of hereditary control problems via approximation technique. Brown Univ. LCDS Tech. Rep. 75-6, Providence, RI, 1975.

C7 H.T. Banks and J.A. Burns: Hereditary control problems: Numerical methods based on averaging approximations. SIAM J. Control and Optimization, Vol. 16, No. 2, pp.169-208, 1978.

C8 H.T. Banks and G.A. Kent: Control of functional differential equations of retarded and neutral type of target sets in functional space. SIAM J. Control and Optimization, Vol. 10, pp. 567-593, 1972.

C9 H.T. Banks and A. Manitius: Application on abstract variational theory to hereditary systems - A survey. IEEE Trans. Automatic Control, AC-19, pp. 524-533, 1974.

C10 H.T. Banks and A. Manitius: Projection series for retarded functional differential equations with applications to optimal control problems. J. Differential Equations, Vol.18, pp. 296-332, 1975.

C11 J.J. Budelis and A.E. Bryson Jr.: Some optimal control results for differential systems. IEEE Trans. on Automatic Control, Vol. AC-15, No. 2, pp. 237-421, 1970.

C12a H.C. Chan and W.R. Perkins: Optimization of time delay systems using parameter imbedding. Automatica, Vol, 9, pp. 257-261, 1973.

C12b A.K. Choudhury: A contribution to the controllability of time lag systems. Int. J. Control, Vol 17, pp. 365-373, 1973.

C13 D.H. Chung and E.B. Lee: Linear optimal systems with time delay. SIAM J. Control, Vol 4, pp. 548-575, 1966.

C14 M.C. Delfour: Numerical solution of the operational Riccati differential equation in the optimal control theory of linear hereditary differential systems with a linear - quadratic cost function. Proc. 1974 IEEE Conference on Decision and Control, pp. 784-790, Phoemix, Ariz, 1974.

C15 M.C. Delfour: Linear hereditary systems and their control. In: Optimal control and its applications. Part II, B.J. Kirby, ed. Springer-Verlag, New York, pp. 92-154, 1977.

C16 M.C. Delfour: The linear quadratic optimal control problem for hereditary differential systems. Theory and Numerical Solution. Applied Mathematics and Optimization, Vol. 3, No. 2/3, pp. 101-162, 1977.

C17 M.C. Delfour and S.K. Mitter: Hereditary differential systems with constant delays. I. General case. J. Differential Equations, Vol. 12, pp. 213-235, 1972.

C18 M.C. Delfour and S.K. Mitter: Hereditary differential equations with constant delays. II. A class of affine systems and the adjoint problem. J. Differential Equations, Vol. 18, pp. 18-28, 1975.

C19 D.H. Eller, J.K. Aggarwal and H.T. Banks: Optimal control of linear time delay systems. IEEE Trans. Aut. Control, AC-14, pp. 678-687, 1969.

C20 S.A. Gracovetsky and M. Vidyasagar: A simple iterative method for sub-optimal control of linear time delay systems with quadratic cost. Int. J. Control, Vol. 16, No. 5, pp. 997-1002, 1972.

C21 R.A. Hess and J.G. Hyde: Suboptimal control of time delay systems. IEEE Trans. Aut. Control, Vol. AC-18, pp. 667-669, 1973.

C22 K. Inoue, K. Ogino and Y. Sawaragi: Sensitivity synthesis of optimal input for parameter identification. Preprint of the 2nd IFAC Symposium on Identification and Process parameter estimation, Praha, Paper No. 9.7, 1970.

C23 K. Inoue, H. Akashi, K. Ogino and Y. Sawaragi: Sensitivity approaches to optimization of linear systems with time delay. Automatica, Vol. 17, No. 6, pp. 671-679, 1971.

C24 M. Jamshidi and M. Malek-Zavarei: Sub-optimal control of linear control systems with time delay. Proc. IEE, Vol. 119, No. 12, pp. 1743-1749, 1972.

C25 G.L. Kharatishvili: Maximum prinsiple in the theory of optimal time delay processes. Dokl. Acad. Sci. USSR 136, pp. 39-42, 1961.

C26 N.N. Krasovskii: Analytic construction of an optimal regulator in a system with time lags. Prickl Mat. Mech. (USSR), Vol. 26, No. 1, 1962.

C27 N. N. Krasovski: Optimal processes in systems with time lags. Proc. 2nd IFAC Conf. (Basel Switzerland, 1963).

C28 H.N. Koivo and E.B. Lee: Controller sysnthesis for linear systems with retarded state and control variables and quadratic cost. Automatica, Vol. 18, pp. 203-208, 1972.

C29 C.S. Lalvani and R.C. Desai: The maximum principle for systems with time delay. Int. Journal of Control, Vol. 18, pp.301-304, 1973.

C30 M.D. Levine: A steepest descent method of synthesizing optimal control programmes. Proc. of I. Mech. E. Convention on Advances in Automatic Control, Nottingham, pp. 29-38, 1965.

C31 D. Mackinnon: Optimal control of systems with pure time delays using variational programming approach. IEEE Trans. Aut. Control, Vol. AC-12, pp. 255-262, 1967.

C32 A. Manitius: Optimal control of time lag systems with quadratic performance indexes. Proc. Fourth IFAC Congress, Warsaw, 1969.

C33 A.W. Olbrot: On degeneracy and related problems for linear constant time-lag systems. Richerche di Automatica, Vol. 3, pp. 203-220, 1972.

C34 A.W. Olbrot: Algebraic criteria of controllability to zero function for linear constant time-lag systems. Control and Cybernetics, Vol. 2, pp. 59-77, 1973.

C35 A. W. Olbrot: Obervability tests for constant time lag systems. Control and Cybernetics, Vol. 4, pp. 71-84, 1975.

C36 R.J.A. Paul and G.C. Legge: Direct sensitivity method of solving boundary value problems in optimal control studies. Proc. IEE (Lund), Vol. 116, No. 2, pp. 273-280, 1969.

C37 V.M. Popov: Pointwise degeneracy of linear time-invariant delay differential equations. Journal of differential equations, Voll 11, pp. 541-561, 1972.

C38 W. H. Ray: The optimal control processes modelled by transfer functions containing pure time delays. Chem. Eng. Sci., Vol. 24, pp. 209-216, 1969.

C39 W.H. Ray and M.A. Soliman: The optimal control processes containing pure time delays-I, Necessary conditions for an optimum. Chem. Energy Sci., Vol 25, pp. 1911-1925, 1970.

C40 Z.V. Rekasius and G.A. Lawrence: Minimum energy control of systems with time delay. IEEE Trans. Aut. Control, Vol. AC-15, pp. 365-368, 1970.

C41 D.W. Ross: Controller design time-lag systems via a quadratic criterion. IEEE Trans. Aut. Control, Vol. AC-16, pp. 664-672, 1971.

C42 Y. Sawaragi,K. Inoue and T. Ohki: Sensitivity synthesis of optimal control under changes of system order. IV. IFAC Congress, Warzawa, Paper No. 68.1, 1969.

C43 Y. Sawaragi, K. Inoue and K. Asai: Synthesis of openloop optimal control with zero terminal constraints. Automatica, Vol. 5, pp. 389-394, 1969.

C44 P. Sannuti: Near optimal design of time delay systems by singular perturbation method. Proc. JACC, Atlanta, USA, pp. 489-496, 1970.

C45 M.A. Soliman and W.H. Ray: Optimal control of multivariable systems with pure time delays. Automatica, Vol. 7, pp. 681-689, 1971.

C46 M.A. Soliman and W.H. Ray: Optimal control for linear quadratic systems having time delays. Int. Jour. Control, Vol.15, No. 4, pp. 609-621, 1972.

C47 A.C.Tsoi: An explicit solution to a class of delay differential equations. Int. Journ. of Control, Vol. 21, pp. 39-48, 1975.

C48 A.C. Tsoi: An explicit solution to a class of delay differential equations. Int. Journ. of Control, Vol.22, pp. 787-799, 1975.

C49 A.C. Tsoi: An explicit solution to a class of functional differential equations. Int. Journ. of Control, Vol.22, pp. 869-875, 1975.

C50 A.C. Tsoi: Recent advances in algebraic system theory of delay differential equations. In: Recent theoretical advances in Control theory. Ed. M. Gregien, Academic Press, pp. 68-129, 1978.

Delay systems and functional differential equations

D1 R. BELLMAN and K.L. COOKE: Differential difference equations, Academic Press, New York, 1963.

D2 R. BELLMAN, R.E. KALABA, and J.A. LOCKETT: Numerical inversion of Laplace transform, American Elsevier, New York, 1966.

D3 K.L. COOKE and S.E. LIST: The numerical solution of integral differential equations with retardation, Tech. Report No. 72-4, Dept. of EE, USCLA, Jan. 1972.

D4 L.E. EL'SGOL'TS and S.B. NORKIN: Introduction to the theory and application of differential equations with deviating arguments, Academic Press, New York, 1973.

D5 L. FOX, D.F. MAYERS, J.R. OCKENDON, and A.B. TAYLOR: On a functional differential equation, Journal of Inst. of Mathematics and Applications, Vol. 8, 1971, pp. 271-307.

D6 R.B. GRAFTON: Periodic solutions of Lienard equations with delay: Some theoretical and numerical results, in Schmitt, K. (Ed.), Delay and functional differential equations and their applications, Academic Press, 1972, pp. 321-334.

D7 E.I. JURY: Theory and application of z-transform methods, John Wiley & Sons, New York, p. 53, 1964.

D8 M. KIELKIEWICZ: Accuracy of Volterra series method for non-linear differential equations, Electronics Letters, Vol. 4, 1968, pp. 584-585.

D9 L. LAPIDUS and W.E. SCHIESSER: Numerical methods for differential systems, Academic Press, N.Y., 1976.

D10 G.R. MORRIS, A. FELDSTEIN, and E.W. BOWEN: The Phragmen-Lindel principle and a class of functional differential equations, in ordinary differential equations, 1971, NRL-NRC Conf. L. Weiss (Ed.), Academic Press, N.Y., 1972, pp. 513-540.

D11 J.R. OCKENDON and A.B. TAYLOR: The dynamics of a current collection system for an electric locomotive, Proc. Roy. Soc., London, A., Vol. 322, pp. 447-468, 1971.

General

G1 G. SANSONE: Orthogonal functions, 1959, N.Y., Interscience.

G2 A.E. TAYLOR: Introduction to functional analysis, 1958, John Wiley & Sons Inc., N.Y.

G3 K. YOSIDA: Functional analysis, 1968, Springer-Verlag, Berlin - Heidelberg - New York.

System identification (General)

I1 K.J. ASTRÖM and P. EYKHOFF: System identification - a survey, Automatica, Vol. 7, 1971, p. 123.

I2 A.V. BALAKRISHNAN and V. PETERKA: Identification in automatic control systems, survey paper in the Proc. of the 4th IFAC Congress, Warsaw, 1969.

I3 S.A. BILLINGS: Identification of nonlinear systems - a survey, Proc. IEE, Vol. 127, Pt-D, No. 6, Nov. 1980, pp. 272-285.

I4 M. CUENOD and A.P. SAGE: Comparison of some methods used for process identification, IFAC Symposium, Identification in automatic control systems, Prague, Paper 1, 1967.

I5 V.K. DESAI and F.W. FAIRMAN: On determining the order of a linear system, Mathematical Biosciences, Vol. 12, 1971, pp. 217-224.

I6 J.E. DIAMESSIS: A new method of determining the parameters of a physical system, Proc. IEEE, Vol. 53, 1965, pp. 205-206.

I7 K. DIEKMANN and H. UNBEHAUEN: Die Identifikation von Mehrgrößensystemen unter Verwendung unterschiedlicher Abtastzeiten, First IASTED Symposium, Modelling, Identification and Control, Davos, Switzerland, Febr. 18-21, 1981.

I8 J. EISENFELD: Remarks on the modulating function method for impulse response identification, IEEE Trans. on Automatic Control, Vol. AC-24, No. 3, June 1979, pp. 498-499.

I9 P. EYKHOFF: System identification, John Wiley, London, 1974.

I10 E. HALFON, H. UNBEHAUEN and C. SCHMID: Model order estimation and system identification theory and application to the modelling of 32 kinetics within the trophogenic zone of a small lake, Ecological Modelling, Vol. 6, 1979, pp. 1-22.

I11 T.C. HSIA: On sampled data approach to parameter identification of continuous linear systems, IEEE Trans. on Automatic Control, Vol. AC-17, No. 2, 1972, pp. 247-249.

I12 T.C. HSIA: A discrete method for parameter identification in linear systems with transportation lags, IEEE Trans. on Aerosp. Electron. Systems, Vol. AES-5, 1969, pp. 236-239.

I13 R.C.K. LEE: Optimal estimation, identification and control, MIT Press, Cambridge, MA, USA, 1964.

I14 L. LJUNG and K. GLOVER: Frequency domain versus time domain methods in system identification, Automatica, Vol. 17, No. 1, 1981, pp. 71-86.

I15 C.S. KUBRUSLY: Distributed parameter system identification - a survey, Int. Journal of Control, Vol. 26, 1977, pp. 509.

I16 G.P. RAO and L. SIVAKUMAR: Identification of deterministic time lag systems, IEEE Trans. on Automatic Control, Vol. AC-21, 1976, pp. 527-529.

I17 H. RAKE: Step response and frequency response methods, Automatica, Vol. 16, 1980, pp. 519-526.

I18 P. ROBERTO GUIDORZI: Invariants and canonical forms for systems structural and parametric identification, Automatica, Vol. 17, No. 1, 1981, pp. 117-133.

I19 A.P. SAGE and J.L. MELSA: System identification, Academic Press, N.Y., 1971.

I20 M. SHINBROT: On the analysis of linear and nonlinear systems, Trans. ASME, 1957, Vol. 79, pp. 547-552.

I21 N.K. SINHA: Estimation of transfer function of continuous system from sampled data, Proc. IEE, May 1972, Vol. 119, No. 5, pp. 612-614.

I22 F.W. SMITH: System Laplace transform estimation from sampled date, IEEE Trans. on Automatic Control, Febr. 1968, Vol. AC-13, pp. 37-44.

I23 M. THOMA: Theorie linearer Regelsysteme, Vieweg-Verlag, 1973.

I24 H. UNBEHAUEN and B. BAUER: Aspects of selection of parameter estimation methods for identification of industrial processes, 4th IFAC Symposium, Tbilisi, 1976.

I25 P. YOUNG: Parameter estimation for continuous-time models - a survey, Automatica, Vol. 17, No. 1, 1981, pp. 23-29.

I26 P. YOUNG, A. JAKEMAN, and R. McMURTRIE: An instrumental variable method for model order identification, Automatica, Vol. 16, 1980, pp. 231-291.

I27 H. UNBEHAUEN and B.G. GÖHRING: Tests for determining model order in parameter estimation, 1974, Automatica, Vol. 10, pp. 233-244.

Poisson moment functional method and related topics

P1 J.E. DIAMESSIS: A new method of determining the parameters of a physical system, Proc. IEEE, 1965, Vol. 53, pp. 205-206.

P2 J.E. DIAMESSIS: On the determination of the parameters of certain nonlinear systems, Proc. IEEE, 1965, Vol. 53, pp. 319-320.

P3 J. EISENFELD: Remarks on the modulating function method for impulse response identification, IEEE Trans. on Automatic Control, Vol. AC-24, No. 3, June 1979, pp. 498-499.

P4 F.W. FAIRMAN and D.W.C. SHEN: Parameter identification for a class of distributed systems, Int. Journal Control, 1970, Vol. 11, No. 6, pp. 929-940.

P5 F.W. FAIRMAN and D.W.C. SHEN: Parameter identification for linear time varying dynamic processes, Proc. IEE, 1970, Vol. 117, No. 10, pp. 2025-2029.

P6 H.S. GREEN and H. MESSEL: On the expansion of functions in terms of their moments, Quart. Appl. Meth., 1954, Vol. 11, pp. 403-409.

P7 A.V. MATHEW and F.W. FAIRMAN: Identification in the presence of initial conditions, IEEE Trans. on Automatic Control, Vol. AC-17, June 1972, pp. 394-396.

P8 A.V. MATHEW and F.W. FAIRMAN: Transfer function matrix identification, IEEE Trans. on circuits and systems, Vol. CAS-21, Sept. 1974, pp. 584-588.

P9 F.J. PERDREAUVILLE and R.E. GOODSON: Identification of systems described by partial differential equations, Journal of Basic, Engineering, Trans. ASME, June 1966, Vol. 88, Series-D, No. 2, pp. 463-468.

P10 G.P. RAO and L. SIVAKUMAR: Identification of deterministic time lag systems, IEEE Trans. on Automatic Control, Vol. AC-21, Aug. 1976, pp. 527-529.

P11 G.P. RAO, D.C. SAHA, T.M. RAO, A. BHAYA and K. AGHORAMURTHY: A microprocessor based system for online parameter identification in continuous dynamical systems, IEEE Trans. on IE, 1982, Vol. IE -29, No. 3, pp. 197-201.

P12 G.P. RAO and L. SIVAKUMAR: Piecewise linear system identification via Walsh functions, 1982, Int. Journal System Science, Vol. 13, No. 5, pp. 525-530.

P13 D.C. SAHA and G.P. RAO: Time domain synthesis via Poisson moment functionals, Int. Journal Control, 1979, Vol. 30, No. 3, pp. 417-426.

P14 D.C. SAHA and G.P. RAO: Identification of lumped linear systems in the presence of unknown initial conditions via Poisson moment functionals, Int. Journal Control, 1980, Vol. 31, No. 4, pp. 637-644.

P15 D.C. SAHA and G.P. RAO: Identification of distributed parameter systems via multidimensional distributions, Proc. IEE, 1980, Vol. 27, Pt. D., CTA, pp. 45-50.

P16 D.C. SAHA and G.P. RAO: Identification of lumped linear time varying parameter systems via Poisson moment functionals, Int. Journal Control, 1980, Vol. 32, No. 4, pp. 709-721.

P17 D.C. SAHA and G.P. RAO: Identification of lumped linear systems in the presence of small unknown time delays via Poisson moment functionals, Int. Journal Control, 1981, Vol. 33, pp. 945-951.

P18 D.C. SAHA and G.P. RAO: A general algorithm for parameter identification in lumped continuous systems - the Poisson moment functional approach, IEEE Trans. on Automatic Control, Febr. 1982, Vol. AC-27, No. 1, pp. 223-225.

P19 D.C. SAHA and G.P. RAO: Transfer function matrix identification in MIMO systems via Poisson moment functionals, Int. Journal Control, 1982, Vol. 35, pp. 727-738.

P20 D.C. SAHA, B.B.P. RAO and G.P. RAO: Structure and parameter identification in linear continuous lumped systems - the Poisson moment functional approach, Int. Journal of Control, 1982, Vol.36. No. 3, pp. 477-491.

P21 L. SCHWARTZ: Mathematics for physical sciences, 1966, Addison Wesley.

P22 A.A. SEIF, A.A. HANAFY and M.F. SAKR: Real time least squares estimation using successive integrations, Information and Control, 1978, Vol. 36, pp. 42-55.

P23 M. SHINBROT: On the analysis of linear and nonlinear systems, Trans. ASME, Vol. 79, April 1957, pp. 547-552.

P24 L. SIVAKUMAR and G.P. RAO: Parameter identification in lumped linear continuous systems in a noisy environment via Kalman-filtered Poisson moment functionals, Int. Journal of Control, March 1982, Vol. 35, No. 3, pp. 509-519.

P25 A.K. MUKHERJEE, D.C. SAHA and G.P. RAO: Identification of large scale distributed parameter systems - Some simplifications in the multidimensional Poisson moment functional (MDPMF) approach, 1983, Int. Journal Systems Science (to appear).

Piecewise constant orthogonal functions and applications

W1 N. AHMED and K.R. RAO: Orthogonal transforms for digital signal processing, Springer-Verlag, New York, 1975.

W2 N. AHMED, H.H. SCHREIBER and P.V. LOPRESTI: On notation and definition of terms related to a class of complete orthogonal functions, IEEE Transactions, Vol. EMC-15, No. 2, May 1973, pp. 75-80.

W3 N. A. ALEXANDRIS: Relations among sequency, axis symmetry, and period of Walsh functions, IEEE Transactions, Vol. IT-17, No. 4, July 1971, pp. 495-497.

W4 K.G. BEAUCHAMP: Walsh functions and their applications, Academic Press, London, 1975.

W5 E.V. BOHN: Estimation of Continuous time linear system parameters from periodic data, Automatica, Vol. 18, No. 1, pp. 27-36, 1982.

W6 N.M. BLACHMAN: Spectral analysis with sinusoids and Walsh functions. IEEE Transactions, Vol. AES-7, No. 5, Sept. 1971, pp. 900-905.

W7 H. BURKHARDT: Ein Beitrag zur Lösung optimaler Steuerungs- und Regelungsprobleme mit Hilfe der Walsh-Transformation, Dr.-Ing. Thesis, Universität Karlsruhe, June 1974.

W8 H. BURKHARDT: Anwendung der Walsh-Transformation auf optimale Steuerungs- und Regelungsprobleme - Eine neue Formulierung der direkten diskreten Lösung. Regelungstechnik, Heft 9, 1975, pp. 294-299.

W9 H. BUTIN: On an ordering of Walsh functions, IEEE Transactions, Vol. C-27, No. 1, Jan. 1978.

W10 J.S. BYRNES and D.A. SWICK: Instant Walsh functions, SIAM Review, Vol. 12, No. 1, Jan. 1970, p. 131.

W11 C.F. CHEN and C.H. HSIAO: Design of piecewise constant gains for optimal control via Walsh functions, IEEE Transactions on Automatic Control, Vol. AC-20, No. 5, Oct. 1975, pp. 596-603.

W12 C.F. CHEN and C.H. HSIAO: A state space approach to Walsh series solution of linear systems, International Journal of Systems Science, Vol. 6, No. 9, Sept. 1975, pp. 833-858.

W13 C.F. CHEN and C.H. HSIAO: Time domain synthesis via Walsh functions, Proceedings IEE, Vol. 122, No. 5, May 1975, pp. 565-570.

W14 C.F. CHEN and C.H. HSIAO: Walsh series analysis in optimal control, International Jou. of Control, Vol. 21, No. 6, June 1975, pp. 881-897.

W15 C.F. CHEN and C.H. HSIAO: A Walsh series direct method for solving variational problems, Journal of Franklin Institute, Vol. 300, No. 4, October 1975, pp. 265-280.

W16 C.F. CHEN, Y.T. TSAY and T.T. WU: Walsh operational matrices for fractional calculus and their applications to distributed systems, Journal of Franklin Institute, Vol. 303, No. 3, March 1977, pp. 267-284.

W17 W.L. CHEN: Block-pulse analysis of scaled systems, International Journal of Systems Science, Vol. 12, No. 7, July 1981, pp. 885-891.

W18 W.L. CHEN and Y.P. SHIH: Analysis and optimization of time-varying linear systems via Walsh functions, International Jou. of Control, Vol. 27, No. 6, June 1978, pp. 917-932.

W19 W.L. CHEN and Y.P. SHIH: Parameter estimation in bilinear systems via Walsh functions, Jou. of Franklin Institute, Vol. 305, No. 5, May 1978, pp. 249-259.

W20 W.L. CHEN and Y.P. SHIH: Shift Walsh matrix and delay differential equations, IEEE Trans. on Automatic Control, Vol. AC-23, No. 6, Dec. 1978, pp. 1023-1028.

W21 D.K. CHENG and J.J. LIU: Walsh-Transform analysis of discrete dyadic-invariant systems, IEEE Transactions, Electromagnetic Compatibility, May 1974, pp. 136-139.

W22 T.M. CHIEN: On representations of Walsh functions, IEEE Transactions, Vol. EMC-17, No. 3, Aug. 1975, pp. 170-176.

W23 M.S. CORRINGTON: Solution of differential and integral equations with Walsh functions, IEEE Transactions on Circuit Theory, Vol. CT-20, No. 5, Sept. 1973, pp. 470-476.

W24 N.J. FINE: On Walsh functions, Transactions Amer. Math. Soc., Vol. 65, 1949, pp. 373-414.

W25a N. GOPALSAMI and B.L. DEEKSHATULU: Comments on 'Design of piecewise constant gains for optimal control via Walsh functions', IEEE Transactions on Automatic Control, Vol. AC-21, No. 4, Aug. 1976, pp. 635.

W25b C.F. CHEN: Reply to the above.

W26 H.F. HARMUTH: Application of Walsh functions in communications, IEEE Spectrum, Vol. 6, No. 11, Nov. 1969, pp. 82-91.

W27 H.F. HARMUTH: Transmission of information by orthogonal functions, Springer-Verlag, Berlin, 1971.

W28 Y.G. JAN and K.M. WONG: Bilinear system identification by block-pulse functions, Jou. Franklin Institute, Vol. 12, No. 5, Nov. 1981, pp. 349-359.

W29 S.C. KAK: Sampling theorem in Walsh-Fourier analysis, Electronics letters, 1970, pp. 447-448, D/12-6-1970.

W30 V.R. KARANAM, P.A. FRICK, and R.R. MOHLER: Bilinear system identification by Walsh functions, IEEE Trans. on Automatic Control, Vol. AC-23, Aug. 1978, pp. 709-713.

W31 R. KITAI: Walsh - to - Fourier spectral conversion for periodic waves, IEEE Transactions, Electromagnetic Compitibility, Nov. 1975, pp. 266-269.

W32 R.W. KOCH and C.R. PAUL: Piecewise-linear-basis-function expansions, Electronics letters, 26. July 1973, Vol. 9, No. 15, pp. 332-334.

W33 C.P. KWONG and C.F. CHEN: The convergence properties of block-pulse series, International Journal of Systems Science, Vol. 12, No. 6, June 1981, pp. 745-751.

W34 M. MAQUSI: On moments and Walsh characteristic functions, IEEE Transactions on Communications, June 1973, pp. 768-770.

W35 P.A. MORETTIN: Walsh function analysis of a certain class of time series, Stochastic processes and their applications, Vol. 4, No. 2, April 1974, pp. 183-194.

W36 K.R. PALANISAMY: Analysis, optimization and identification of lumped continuous systems with time delays via Walsh functions, Ph. D. Thesis, Dept. of Electrical Engineering, I.I.T., Kharagpur, 721 303, India, August 1980.

W37a K.R. PALANISAMY and G. P. RAO: Minimum energy control of time delay systems via Walsh functions, 1983, Optimal Control Applications and Methods (to appear).

W37b K.R. PALANISAMY: Analysis and optimal control of linear systems via single term Walsh series approach, International Jou. of Systems Science, Vol. 12, No. 4, 1981, pp. 443-454.

W38 K.R. PALANISAMY and D.K. BHATTACHARYA: System identification via block-pulse functions, International Jou. of Systems Science, Vol. 12, No. 5, 1981, pp. 643-647.

W39 P.N. PARASKEVOPOULOS and A.C. BOUNAS: Distributed parameter system identification via Walsh functions, International Jou. of Systems Science, Vol. 9, No. 1, 1978, pp. 75-83.

W40 C.R. PAUL and R.W. KOCH: On piecewise-linear basis functions and piecewise-linear signal expansions, IEEE Transactions, Vol. ASSP-22, No. 4, August 1974, pp. 263-268.

W41 J. PEARL: Application of Walsh transform to statistical analysis, IEEE Transactions, Vol. SMC-1, No. 2, April 1971, pp. 111-119.

W42 G.P. RAO and K.R. PALANISAMY: A new operational matrix for delay via Walsh functions and some aspects of its algebra and applications, National Systems Conference, NSC-78, PAU Ludhiana (India), Sept. 1978.

W43a G.P. RAO and K.R. PALANISAMY: Optimal Control of time-delay systems via Walsh functions, 9th IFIP Conference on 'optimization techniques', Polish Academy of Sciences, Systems Research Institute, Warsaw, Poland, Sept. 1979.

W43b G.P. RAO and K.R. PALANISAMY: Walsh stretch matrices and functional differential equations, IEEE Trans., Vol. AC-21, No. 1, February 1982, pp. 272-276.

W44 G.P. RAO and K.R. PALANISAMY: Improved algorithms for parameter identification in lumped continuous systems via Walsh functions, Proc. IEE (Lond.), Pt-D, CTA, Vol. 130, No.1, Jan. 1983, pp. 9-16.

W45 G.P. RAO, K.R. PALANISAMY and T. SRINIVASAN: Extension of computation beyond the limit of initial normal interval in Walsh series analysis of dynamical systems, IEEE Trans. on Automatic Control, Vol. AC-25, No. 2, April 1980, pp. 316-319.

W46 G.P. RAO and L. SIVAKUMAR: System identification via Walsh functions, Proc. IEE, Vol. 122, No. 10, October 1975, pp. 1160-1161.

W47 G.P. RAO and L. SIVAKUMAR: Identification of time-lag systems via Walsh functions, IEEE Transactions on Automatic Control, Vol. AC-24, No. 5, October 1979, pp. 806-808.

W48 G.P. RAO and L. SIVAKUMAR: Transfer function matrix identification in MIMO systems via Walsh functions, Proc. IEEE, Vol. 69, No. 4, April 1981, pp. 465-466.

W49 G.P. RAO and L. SIVAKUMAR: Order and parameter identification in continuous linear systems via Walsh functions, Proc. IEEE, Vol. 70, No. 7, July 1982, pp. 764-766.

W50 G.P. RAO and L. SIVAKUMAR: Piecewise linear system identification via Walsh functions, Int. Jou. Systems Science, Vo. 13, No. 5, 1982, pp. 525-530.

W51 G.P. RAO and T. SRINIVASAN: Remarks on "Authors' reply" to 'Comments on design of piecewise constant gains for optimal control via Walsh functions', IEEE Transactions on Automatic Control, Vol. AC-23, No. 4, Aug. 1978, pp. 762-763.

W52 G.P. RAO and T. SRINIVASAN: Solution of certain nonlinear functional equations via block-pulse functions, Proceedings of the 5th National Systems Conference, NSC-78, PAU, Ludhiana (India), September 1978, pp. 1-287-1-290.

W53 G.P. RAO and T. SRINIVASAN: Analysis and Synthesis of dynamic systems containing time-delays via block-pulse functions, Proc. IEE, Vol. 125, No. 9, October 1978, pp. 1064-1068.

W54 G.P. RAO and T. SRINIVASAN: An optimal method of solving differential equations characterizing the dynamics of a current collection system for an electric locomotive, Journal of the Inst. of Math. and its Applications, Vol. 25, No. 4, June 1980, pp. 329-342.

W55 G.P. RAO and T. SRINIVASAN: Multi-dimensional blockpulse functions and their use in the study of distributed parameter systems. Int. Journal Systems Science, 1980, Vol. 11, No. 6, pp. 689-708.

W56 V.A. ROMANOV and V.A. SEMERAN: Algorithm for identifying the dynamic characteristics of objects by means of orthogonal Walsh functions, Automation and Remote Control, Vol. 34, 1973, pp. 601-607.

W57 P. SANNUTI: Analysis and synthesis of dynamic systems via block-pulse functions, Proceedings IEE, Vol. 124, No. 6, June 1977, pp. 569-571.

W58 H.H. SCHREIBER: Bandwidth requirements for Walsh functions, IEEE Transactions, Vol. II-16, July 1970, pp. 491-492.

W59 L.S. SHIEH and R.E. YATES: Solving inverse Laplace transform, linear and nonlinear state equations, using block pulse functions, Comput. and Elect. Engineering, Vol. 6, 1979, pp. 3-17.

W60 L.S. SHIEH, C.K. YEUNG, and B.C. McINNIS: Solution of state space equations via block-pulse functions, Int. Journal Control, 1978, Vol. 28, No. 3, pp. 383-392.

W61 Y.P. SHIH and J.Y. HAN: Double Walsh series solution of first order partial differential equations, Int. Journal Systems Science, 1978, Vol. 9, No. 5, pp. 569-578.

W62 Y.P. SHIH and W.K. CHIA: Piecewise constant solutions of integral equations via Walsh functions, Journal of the Chinese Institute of Engineers, No. 1, Vol. 1, pp. 81-85 (1978).

W63 Y.M. SHIH: Block pulse function analysis of time varying and nonlinear networks, Journal of the Chinese Institute of Engineers, Vol. 1, No. 2, pp. 43-52, 1978.

W64 M.S.P. SINHA, V.S. RAJAMANI and A.K. SINHA: Identification of nonlinear distributed systems using Walsh functions, Int. Jou. Control, 1980, Vol. 32, No. 4, pp. 669-676.

W65 L. SIVAKUMAR: Some aspects of system identification leading to Walsh function approach, Ph. D. Thesis, Dept. of Electrical Engineering, I.I.T., Kharagpur 721 302, India, September 1978.

W66 T. SRINIVASAN: Analysis of dynamical systems via block-pulse functions, Ph. D. Thesis, Dept. of Electrical Engineering, I.I.T., Kharagpur 721 302, India, December 1979.

W67 P. STAVROULAKIS and S. TZAFESTAS: A Walsh series approach to time-delay control system observer design, Int. Journal Systems Science, Vol. 9, No. 3, 1978, pp. 287-299.

W69 D.A. SWICK: Walsh function generation, IEEE Transactions, Vol. IT-15, No. 1, Jan. 1969, pp. 167.

W70 S. TZAFESTAS: Walsh series approach to lumped and distributed system identification. Journal of Franklin Institute, Vol. 305, No. 4, April 1978, pp. 199-220.

W71 S. TZAFESTAS and N. CHRYSOCHOIDES: Nuclear reactor control using Walsh function variational synthesis, Nuclear Science and Engineering, Vol. 62, 1977, pp. 763-770.

W72 S. TZAFESTAS and P. STAVROULAKIS: Walsh series approach to observer and filter design in optimal control systems, International Journal of Control, Vol. 26, No. 5, Nov. 1977, pp. 721-736.

W73 TIU LE VAN, LE DINH CHON TAM, and NOEL VAN HOUTTE: On direct algebraic solutions of linear differential equations using Walsh transforms, IEEE Transactions, Vol. CAS-22, No. 5, May 1975, pp. 419-422.

W74 S. VENKATESAM and G.P. RAO: A Walsh spectral analyser, Int. Journal Electronics, 1979, Vol. 46, No. 4, pp. 413-415.

W75 J.L. WALSH: A closed set of normal orthogonal functions, Amer. J. Math., Vol. 45, 1923, pp. 5-24.

W76 C.K. YUEN: An algorithm for computing correlation functions of Walsh functions, IEEE Transactions, Vol. EMC-17, No. 3, August 1975, pp. 177-180.

INDEX

I

J

K

L

M

N

O

P

R

S

Lecture Notes in Control and Information Sciences

Edited by A. V. Balakrishnan and M. Thoma